1 MONTH OF
FREE
READING

at
www.ForgottenBooks.com

By purchasing this book you are eligible for one month membership to ForgottenBooks.com, giving you unlimited access to our entire collection of over 1,000,000 titles via our web site and mobile apps.

To claim your free month visit:

www.forgottenbooks.com/free904554

ISBN 978-0-266-88557-3
PIBN 10904554

 **United States Department of Commerce**
Technology Administration
National Institute of Standards and Technology

NIST Technical Note 1500-3
Materials Reliability Series

High-Energy, Transmission X-Ray Diffraction for Monitoring Turbine-Blade Solidifaction

Dale W. Fitting
William P. Dubé
Thomas A. Siewert

United States Department of Commerce
Technology Administration
National Institute of Standards and Technology

NIST Technical Note 1500-5
Materials Reliability Series

NIST Technical Note 1500-3
Materials Reliability Series

High-Energy, Transmission X-Ray Diffraction for Monitoring Turbine-Blade Solidifaction

Dale W. Fitting
William P. Dubé
Thomas A. Siewert

Materials Reliability Division
Materials Science and Engineering Laboratory
National Institute of Standards and Technology
325 Broadway
Boulder, Colorado 80303-3328

June 1998

U.S. DEPARTMENT OF COMMERCE, William M. Daley, Secretary
TECHNOLOGY ADMINISTRATION, Gary R. Bachula, Acting Under Secretary for Technology
NATIONAL INSTITUTE OF STANDARDS AND TECHNOLOGY, Raymond G. Kammer, Director

National Institute of Standards and Technology Technical Note
Natl. Inst. Stand. Technol., Tech. Note 1500-3, 156 pages (June 1998)
CODEN:NTNOEF

U.S. GOVERNMENT PRINTING OFFICE
WASHINGTON: 1998

CONTENTS

FOREWORD

The Materials Reliability Series of NIST Technical Notes are reports covering significant research accomplishments of the Materials Reliability Division. The Division develops measurement technologies that enable the producers and users of materials to improve the quality and reliability of their products. Measurement technologies are developed for process control to improve the quality and consistency of materials, for nondestructive evaluation to assure quality of finished materials and products, and for materials evaluation to assure reliable performance. Within these broad areas of measurement technology, the Division has focused its resources on three research themes:

- Intelligent Processing of Materials—To develop on-line sensors for measuring the materials' characteristics and/or processing conditions needed for real-time process control.

- Ultrasonic Characterization of Materials—To develop ultrasonic measurements for characterizing internal geometries of materials, such as defects, microstructures, and lattice distortions.

- Micrometer-Scale Measurements for Materials Evaluation—To develop measurement techniques for evaluating the mechanical, thermal, and magnetic behavior of thin films and coatings at the appropriate size scale.

This report is the third in the Materials Reliability Series. It covers research on high-energy transmission X-ray diffraction for monitoring turbine-blade solidification. Previous reports in this series are:

Technical Note 1500-1   Tensile Testing of Thin Films: Techniques and Results, by D.T. Read, 1997

Technical Note 1500-2   Procedures for the Electron-Beam Moiré Technique, by E.R. Drexler, 1998

HIGH-ENERGY, TRANSMISSION X-RAY DIFFRACTION FOR MONITORING TURBINE-BLADE SOLIDIFICATION

Dale W. Fitting, William P. Dubé, and Thomas A. Siewert
Materials Reliability Division
Materials Science and Engineering Laboratory
National Institute of Standards and Technology
Boulder, Colorado 80303

We developed a nondestructive x-ray technique to monitor the solidification of single-crystal turbine blade castings. X-ray energies in the 150 keV to 300 keV range have sufficient energy for transmission x-ray diffraction to be performed on a 12 mm thick nickel-alloy specimen. We obtained diffraction images even though the x-ray path (over 1 m) through a directional solidification furnace included glass furnace ports, molybdenum heater windings, an aluminum oxide heater-coil support, and mold material. The method was capable of sensing changes in the physical state of the casting (liquid, dendrite, solid) and measuring the solid fraction in the mushy zone.

Keywords: diffraction, directional solidification, investment casting, single-crystal, x-ray

EXECUTIVE SUMMARY

We developed a technique based on transmission x-ray diffraction (XRD) to study the solidification of a single-crystal turbine blade casting within its mold. High-energy x-rays penetrate through material surrounding the casting and produce a distinctive diffraction pattern which clearly indicates whether the sampled region is liquid or solid. A real-time transmission Laue x-ray image of the casting shows an ordered pattern of x-ray scattering (diffraction spots) from the solid and a diffuse ring of scattering from the liquid. The dramatically different spatial pattern provides a high-contrast, unequivocal spatial discrimination of the physical state of the alloy.

Proof-of-concept experiments were performed on aluminum and copper melted in a gradient furnace. Laue images and transmission energy spectra proved to be highly sensitive means for identifying the presence of liquid or solid material in the casting.

We have successfully performed real-time transmission XRD on a casting of N5 nickel-alloy in an industrial furnace for casting turbine blades. The high energies (up to 320 keV), permitted transmission XRD to be performed on a specimen 12 mm thick. XRD images were obtained even though the x-ray path through the furnace included 20 mm of borosilicate glass (furnace ports), 3.2 mm of molybdenum (furnace hot-zone windings), 9 mm of aluminum oxide (hot-zone coil support), and 12.8 mm of mold material (refractory oxides). The distance from x-ray source to imager was 1180 mm. Motion-control stages permitted scanning of both the x-ray source and the detector to follow the casting during withdrawal and to probe the mushy zone. The x-ray diffraction spot intensity was high when the probing x-ray beam was directed into the solid, and decreased when the x-ray beam was scanned vertically into the molten alloy. Plots of diffraction spot intensity versus vertical position (increasing temperature) in the casting correlate well with Lever-law model predictions of fraction solid versus temperature.

Accomplishments

- We developed high-energy, transmission x-ray diffraction as a noninvasive means for locating and characterizing the liquid-dendrite-solid region in a single-crystal investment casting during directional solidification.

- We demonstrated the x-ray diffraction sensing technology on a mold-encased nickel alloy (N5) specimen in a turbine-blade casting furnace.

- We developed an analytical model for high-energy transmission diffraction and used it to optimize the diffraction system for specific mold/sample configurations.

- We received a patent on the transmission diffraction technique for monitoring solidification of a casting within its mold.

1. INTRODUCTION

The NIST Consortium on Casting of Aerospace Alloys, an industry/government/university team, was organized in 1994 to improve quality and reduce cost of aerospace castings through advances in materials science. The Consortium identified several areas where developments in technology would substantially improve the investment casting process. In particular, no sensor had been capable of locating the liquid-dendrite-solid region in the harsh environment of a turbine-blade casting furnace (high vacuum, high temperatures, and strong radio frequency fields).

A variety of sensing technologies (x-ray imaging, ultrasonics) have been used to study metal castings [1-5]. Yet, few techniques are suitable for process sensing when the metal is encased in a thick ceramic mold with irregular interior and exterior geometry. We have developed a transmission x-ray diffraction sensor capable of monitoring the position, shape, and extent of the liquid-dendrite-solid region (mushy zone) in single-crystal turbine-blade castings.

The technique is capable of sensing the physical state of the casting even though the casting is enclosed in a thick ceramic mold and contained within a vacuum furnace. The high contrast (many times greater than that for a conventional x-ray imaging technique) of the XRD sensor is achieved because of the strikingly different x-ray diffraction patterns from a solid versus a liquid.

An x-ray beam directed into a crystalline material will diffract sharply from the lattice planes. The spatial distribution of the diffracted x-rays depends on the spacing of the crystalline planes and the orientation of the x-ray beam with respect to these planes. However, the emergence of any geometric pattern in the scattering indicates the presence of an ordered structure. X-ray scattering from a liquid is relatively amorphous, with only a single broad peak in its radial distribution. Thus, the presence or absence of a sharp geometric pattern of diffracted x-rays forms the basis for a sensing method for demarcating the liquid-solid boundary in a single-crystal casting. In addition to sensing the position of the solidification front, the XRD method has been used to determine the extent and shape of the mushy zone in a casting of nickel superalloy.

Analytical x-ray diffraction is conducted in the reflection geometry using x-ray energies in the 5 to 17 keV range. This technique returns crystal structure information from only the surface and several micrometers beneath it. Since the penetration depth of low-energy x-rays is shallow, traditional x-ray diffraction is thus unable to probe the interior of thicker structures. Synchrotron radiation, producing x-ray energies of 100 keV and higher, permits the study of specimens up to a few-millimeters thick [6,7]. In this paper, we show that melting and solidification of a centimeter-thick metallic specimen within a mold can be monitored by using x-rays of higher energies, 150 to 300 keV, and a transmission configuration.

Using x-ray diffraction to study metal solidification and phase changes is not new [8-14]. However, most of this prior research used very thin (a few millimimeters at most) specimens, furnaces with low-attenuation x-ray windows (beryllium, graphite, or polyimide), and low (less than 50 keV) x-ray energies. Work by Green [15] extended x-ray diffraction investigations to energies exceeding 150 keV. His flash x-ray diffraction system also provided the facility for studying structural changes during dynamic events, such as crystal growth. Others, including Kopinek, et al. [16] and Bechtoldt, et al. [17] have employed high-energy XRD for studies of stress and texture in thick (up to 12.7 mm) steel specimens.

Probing the interior of a casting requires a transmission configuration. X-ray energies of over 150 keV are needed to penetrate the refractory oxide mold (5 to 10 mm wall thickness) and casting specimen (1 to 10 mm thick). These x-rays have energy sufficient to penetrate the furnace components, mold, and alloy specimen, while retaining a moderate cross section for diffraction. The work reported herein shows that melting and solidification of a thick metal specimen can be detected nondestructively from the spatial distribution or energy spectrum of diffracted x-rays.

Solidification sensing is carried out in the following manner. An x-ray beam of small diameter is directed into the material being cast. Detection of only spots indicates a fully solid structure, detection of only a diffuse ring indicates a fully liquid region, and detection of a combination of spots and diffuse ring indicates a mixture of liquid and solid. The relative intensities of the spots and ring are proportional to the fraction of solid and fraction of liquid (respectively) present at any location. Scanning the x-ray source and detector to move the region probed by the x-ray beam may be used to locate the mushy zone in an alloy casting.

2. THEORY

An understanding of x-ray physics is required in sections of this report, Therefore, we present a brief tutorial. Additional information may be found in references 18 through 20.

2.1 Tutorial on X-Rays and Their Interactions with Matter

X-rays are electromagnetic waves with an energy in the 1 keV to 50 MeV range. The wavelength (λ), frequency (f), and speed (c) are related through the expression

$$c = \lambda f .$$ (1)

The energy carried by an x-ray is

$$E = \frac{h c}{\lambda} ,$$ (2)

where
   h is Planck's constant (6.626076×10^{-34} J·s), and
   c is the speed of light (2.997925×10^8 m/s, in vacuum).

With λ in nm and E in keV, the expression becomes

$$E = \frac{1.239842}{\lambda} .$$ (3)

For wavelengths in the few nanometer range (atomic dimensions), the corresponding energy of an x-ray photon is a few kiloelectronvolts (1 kev = 1.602177×10^{-16} J).

2.1.1 X-Ray Interactions

Below energies of 1.02 MeV, x-ray photons interact primarily with the electrons of atoms, by photoelectric absorption, Compton scattering, or elastic scattering. In a photoelectric interaction the incident x-ray energy is totally transferred to an electron, usually an inner-shell electron. If the energy imparted during the interaction is larger than the binding energy of that electron, the electron is ejected from the atom. The kinetic energy of the electron is equal to the incident x-ray energy minus the binding energy of the electron. The resulting electronic structure of the atom is unstable. Stability is restored when the vacancy is filled by a more loosely bound (outer shell) electron. During this transition, a photon, with energy equal to the difference in binding energies of the inner and outer shells is released from the atom. This process is termed fluorescence radiation. During some outer to inner shell transitions, the energy is imparted to an electron,

4

rather than being released as a photon. The ejected electron is called an Auger electron. The energy of the emitted radiation (photon or electron) is characteristic of the atom, and may be used identify the element.

X-ray interactions with loosely bound, outer shell electrons occur by Compton scattering. The electron is ejected and the incident photon (with decreased energy and momentum, and increased wavelength) is scattered incoherently. The phase of the scattered x-ray is not related to that of the incident x-ray; it is random. Incoherent Compton scattering is usually a nuisance in x-ray imaging or diffraction studies—it contributes to the background and carries little useful information.

An x-ray photon may also be elastically scattered from an electron in an atom. The electric field of the incident x-ray causes the electron to oscillate. The moving charge reradiates the electromagnetic wave with no energy loss. Because the phase of the incident and reradiated electromagnetic field are directly related the process is termed coherent scattering. It is the coherence of this process which generates x-ray diffraction.

2.1.2 Attenuation Coefficients

Interactions of a narrow beam of monoenergetic x-rays with matter may be characterized with a total mass attenuation coefficient (μ/ρ). The intensity of x-rays transmitted (at perpendicular incidence) through a slab of material is given by

$$I = I_0\, e^{-\frac{\mu}{\rho}\rho x}, \tag{4}$$

where
I_0 is the incident x-ray intensity,
I is the transmitted x-ray intensity,
μ/ρ is the mass attenuation coefficient (a function of x-ray energy),
ρ is the density of the material, and
x is the thickness of the material.

Figure 1 illustrates the behavior of transmitted intensity by plotting fractional transmission I/I_0 as a function of the material thickness. Raising x-ray energy increases transmission, because the attenuation coefficient decreases with energy.

A linear attenuation coefficient (μ), defined as the product of the mass attenuation coefficient and the density, is useful for comparing the relative transmission through different materials of the same thickness. The reciprocal of the linear attenuation coefficient (that thickness of material which reduces the intensity to $1/e$) is often used as a measure of the effective penetration depth for x-rays. Plots of $1/\mu$ for several materials are shown in Figure 2.

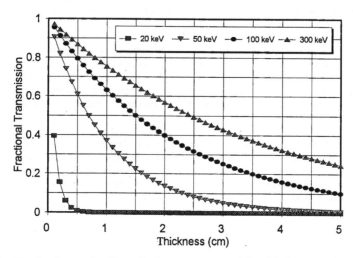

Figure 1. Fractional transmitted intensity decreases exponentially with thickness (shown here for nickel); transmission depends on photon energy.

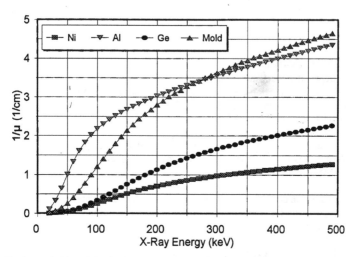

Figure 2. Reciprocal, $1/\mu$, of the linear attenuation coefficient, μ, as a function of x-ray energy. The physical meaning of $1/\mu$ is the effective penetration depth.

The total mass attenuation coefficient may be expressed in terms of a sum of partial mass attenuation coefficients, which reveal the contributions by the photoelectric (*PE*), Compton (*Comp*), and coherent (*coh*) processes.

$$\left(\frac{\mu}{\rho}\right)_{total} = \left(\frac{\mu}{\rho}\right)_{PE} + \left(\frac{\mu}{\rho}\right)_{Comp} + \left(\frac{\mu}{\rho}\right)_{coh} . \tag{5}$$

Partial and total mass attenuation coefficients for all elements and a variety of mixtures and compounds have been tabulated [21]. Software is also available for calculating attenuation coefficients [22]. Plots of attenuation coefficients for several materials as a function of energy are shown in Figures 3 through 6. The attenuation coefficients are significantly larger for nickel and tungsten having higher atomic number and density, than for aluminum or the mold material (a mixture of refractory oxides—alumina, zirconia, and silica). The abrupt change in attenuation coefficient at particular energies (refer to the plot for tungsten) is termed an absorption edge. It represents the increased probability of interaction of an x-ray with a material when the x-ray energy is sufficient to eject a bound electron by the photoelectric effect. Absorption edges for aluminum, nickel and the mold material occur at energies lower than 20 keV. Partial and total mass attenuation coefficients for many of the materials used in our studies of x-ray diffraction may be found in the appendices.

The dominant interactions of x-rays with matter are the photoelectric process at low energies and Compton scattering at higher energies. Although coherent scattering is not a large contributor to the total attenuation, this interaction does occur at high energies.

2.2 X-Ray Production

X-rays are produced when a moving charged particle interacts with a target material. In a conventional x-ray tube, electrons ejected from a heated filament (cathode) are accelerated by a high voltage toward a metal anode. If an incident electron ejects an electron from an atom in the anode, the vacancy in the electron shell is filled by an electron from an outer shell. During the transition, an x-ray of specific energy is produced. The energy of the x-ray is characterisitic of the shell transition within the anode material. When the innermost shell electrons are ejected, K-characteristic x-rays are produced. $K\alpha_1$, $K\alpha_2$, and $K\beta_1$ x-rays of discrete energies are produced when electrons from the L3, L1, and M3 shells fill a K-shell vacancy. The intensity (I_{char}) of characteristic x-rays is linearly related to the current (i) passing through the tube and the difference between the tube potential (V) and the threshold value (V_{thresh}) of the potential for producing a specific characterisitic x-ray line.

$$I_{char} = C_1 \, i \, (V - V_{thresh})^n , \tag{6}$$

where C_1 is a constant, and n is approximately 1.5.

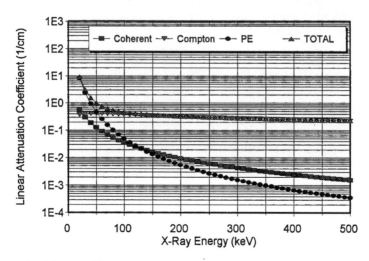

Figure 3.   Partial and total linear attenuation coefficients for aluminum (Al) as a function of x-ray energy.

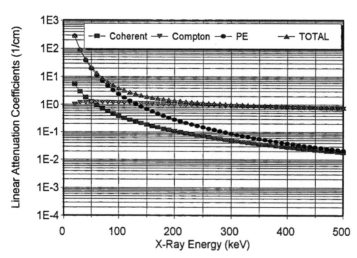

Figure 4.   Partial and total linear attenuation coefficients for nickel (Ni) as a function of x-ray energy.

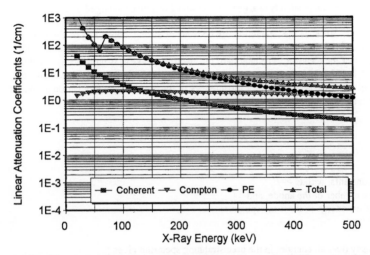

Figure 5.   Partial and total linear attenuation coefficients for tungsten (W) as a function of x-ray energy.

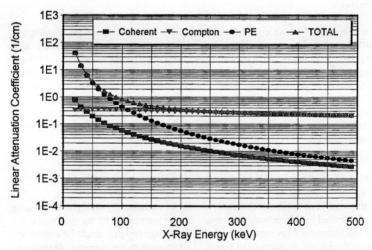

Figure 6.   Partial and total linear attenuation coefficients for the mold material (mixture of alumina, zirconia, and silica) used to make single-crystal turbine blade castings.

Electrons accelerated in the x-ray tube can also interact with the nuclei of anode atoms. The attractive force between the positively charged nucleus and a negatively charged electron can slow the electron. The decelerating charge emits electromagnetic radiation in the form of bremstrahlung x-rays. Because the interaction can be weak or strong, depending on the proximity of the electron to the nucleus, the x-ray energy can range continuously from a small value to the total kinetic energy of the electron. The energy of the electrons, in keV, is numerically equal to the tube potential, in kV (100 keV maximum energy with a 100 kV tube potential). Kulenkampf [23] developed an empirical relationship for the intensity of the bremstrahlung radiation produced at a particular wavelength. We have rewritten the relationship here, in terms of energy (E).

$$ I(E) = C\ Z\ E^2\ (E_{max} - E) + B\ Z^2\ E^2\ , \tag{7} $$

where
C and B are constants, with $C >> B$,
Z is the atomic number of the anode, and
E_{max} is the highest energy in the x-ray spectrum.

Integrating intensity over all energies in the bremstrahlung spectrum gives

$$ I_b = C_2\ i\ Z\ V^2\ , \tag{8} $$

where C_2 is a constant.

The spectrum of energies from an x-ray tube is a composite of these interactions, yielding a continuous spectrum produced by bremstrahlung interactions, overlaid with characteristic x-ray peaks. Increasing the tube potential (V) dramatically increases the intensity of both characteristic and bremstrahlung radiation. X-ray intensity increases linearly with tube current (i). Figure 7 is an x-ray spectrum, measured experimentally, from a tungsten-target tube, operated at a tube potential of 75 kV. Significant features are the smooth bremstrahlung spectrum with a maximum energy of 75 keV, the intense characteristic x-ray peaks (at 58.0, 59.3, 67.2 keV) from the tungsten anode, and a maximum intensity in the bremstrahlung spectrum at approximately one-third of the maximum energy.

2.3 Radiographic Imaging

The interior structure of optically opaque materials may be examined by radiographic imaging. The specimen is illuminated by a large-area x-ray field. X-rays transmitted by the specimen are recorded on a two-dimensional x-ray sensitive medium. The image receptor might be a sheet of film or an electronic device which responds to x-rays. Transmission by the specimen depends on the attenuation coefficient, density, and thickness. Since the attenuation coefficient is a function of energy, the choice of x-ray tube voltage will affect the contrast of the radiographic image.

10

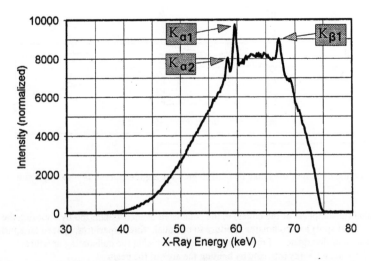

Figure 7.   Spectrum of x-rays from a tube with a tungsten anode. Tube potential = 75 kV, tube current = 0.2 mA, source-to-detector distance = 960 mm.

2.4 Collimation

Certain x-ray procedures (such as diffraction) require that the x-ray beam is restricted to a very narrow beam. Collimators perform this task. X-ray collimators generally do not make the rays parallel (as do optical collimators) but simply restrict the emerging rays to a narrow range of angles [24]. Apertures or slits in a very attenuating material (lead, tungsten) are commonly used. Figure 8 shows the geometry of a set of collimating apertures. The maximum divergence angle is given by

$$\tan (\beta/2) = \frac{d/2}{u/2} \,, \tag{9}$$

where the angles are defined in Fig. 8. Since the angle β is small,

$$\beta \sim \frac{2d}{u} \,, \tag{10}$$

where β is in radians.

11

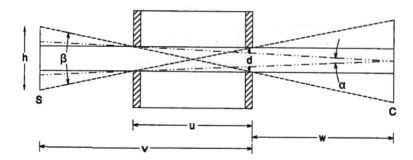

Figure 8.   Collimating apertures may be used to limit the angular spread of an x-ray beam.

Penumbra (unsharpness at the edges) in the collimated beam is minimized by choosing the source size (focal spot) and collimator aperture to be equal. Small apertures, spaced far apart will restrict the angular divergence of the beam. However, reducing the collimating apertures decreases the available x-ray intensity by limiting the area of the beam.

2.5 X-Ray Detectors

X-ray imaging or measurements require a device for measuring the intensity of an x-ray beam. Silver bromide grains in film undergo a change in energy state when exposed to x-rays. Chemical processing causes silver in these affected film grains to precipitate, producing a dark spot on the film. Film is able to record the two-dimensional distribution of x-ray intensities, and so is useful for radiographic imaging. Two other types of detectors were used in our research—a real time imaging device and an energy-sensitive germanium detector. Both are discussed in detail later in this report.

2.6 X-Ray Diffraction (XRD)

Constructive interference of coherently scattered x-rays produces diffraction. Since coherent scattering is an interaction of an electromagnetic wave with atomic electrons, let us begin by examining the scattering from a single electron. Thomson described the scattering of an unpolarized electromagnetic wave from a single electron with the equation [18]

$$I_p = I_0 \; \frac{e^4}{R^2 \, m^2 \, c^4} \; \frac{[1+\cos^2(2\theta)]}{2} \; (\frac{1}{4\pi \; \epsilon_0})^2 \;, \tag{11}$$

12

where
  I_p is the scattered intensity,
  e is the charge of an electron,
  R is the distance from the electron where the intensity is measured,
  m is the mass of an electron,
  c is the speed of light in a vacuum,
  2θ is the angle between the incident and scattered directions, and
  ϵ_0 the dielectric constant of free space.

Evaluating the constant term gives

$$\frac{e^4}{m^2 c^4} \left(\frac{1}{4\pi \epsilon_0}\right)^2 = 7.88 \times 10^{-30} \ [m^2] \ . \tag{12}$$

However, in 1 mg of material there are over 10^{20} electrons, and at $R = 10^{-1}$ m,

$$\frac{I_p}{I_0} \approx 10^{-7} \ . \tag{13}$$

Although this ratio seems small, it is easily detected.

Next, consider scattering from the cloud of electrons surrounding an atom. Figure 9 shows a plane electromagnetic wave incident upon an atom. Suppose coherent scattering from two of the electrons occurs. Because the path lengths travelled by the electromagnetic waves differ, the waves scattered to the point of observation will differ in phase. Interference of the scattered waves will depend upon the location of the electrons (spatial probability density distribution), on the scattering angle (2θ), and on the wavelength of the incident wave. An atomic scattering factor (f) which describes this interference, may be calculated for a particular atom or ion using a quantum mechanical approach. The atomic scattering factor (sometimes called the form factor) is

$$f = \frac{E_{atom}}{E_e} \ , \tag{14}$$

where
  E_{atom} is the electric field scattered from an atom, and
  E_e is the field scattered from a single electron (that is Thomson scattering).

In general, f will be a complex number [25], where

$$Re(f) = f_0 + f' + f_{NT} \ , \quad f' = f_1 + f_{rel} - Z \ , \tag{15}$$

13

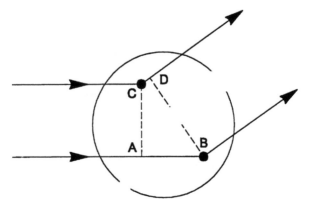

Figure 9.   Diagram demonstrating coherent scattering of an electromagnetic wave, incident from the left, on two electrons in an atom.

and

$$Im(f) = f'' = f_2 = \frac{E \,\mu_{PE}\, u\, A}{2\,h\,c\,r_e} \,,$$ (16)

where
  E is the photon energy,
  f_0 is the coherent scattering factor,
  f_{NT} is the nuclear Thomson factor,
  f_{rel} is the relativistic correction factor,
  u is the atomic mass unit (1/12 the mass of an atom of ^{12}C),
  A is the relative atomic mass of the material, and
  r_e is the radius of an electron.

The terms f' and f'' are called the anomalous dispersion corrections. They are especially significant near the absorption edges of the scattering atom. f_1 and f_2 are functions of energy and atomic number, and f_0 is a function of the momentum transfer $(4\pi \sin(\theta)/\lambda)$ and energy. f_{NT} and f_{rel} are small, negative numbers. Equation (16) also shows the imaginary part of the dispersion correction is related to the photoelectric attenuation coefficient. The real and imaginary components of f have been calculated, and are tabulated [25]. Plots of $f_1(E)$ and $f_2(E)$ for nickel are shown in Figures 10 and 11, respectively.

In some of our models we use an empirical fit [26]

$$f_0 = \sum_{i=1}^{5} a_i \, \exp[-b_i \, (\frac{\sin\theta}{\lambda})^2] + c$$ (17)

14

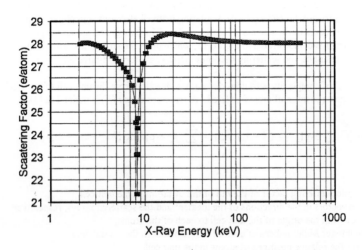

Figure 10. Plot of f_1, for nickel, as a function of energy. f_1 approximates the real part of f in the forward direction.

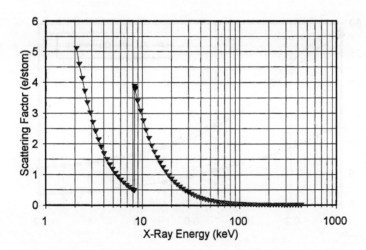

Figure 11. Plot of f_2, for nickel, as a function of energy. f_2 can be related to the partial attenuation coefficient for photoelectric interactions.

15

to the coherent scattering factor. The constants a_i, b_i, and c depend upon the atom. Plots of atomic scattering factor for several different elements are shown in Figure 12, as a function of $(\sin\theta/\lambda)$. In the forward direction $(\theta = 0)$, scattering from all electrons is in phase, so the scattering factor is equal to the number of electrons (Z). The energy and angular dependence of the scattering factor is demonstrated in Figure 13, which plots the atomic scattering factor for nickel. For high energies, the scattering becomes increasingly forward-directed.

The next level of complexity is to consider x-ray diffraction from the unit cell of a crystal, where the structure (position of atoms) is fixed. A structure factor (F) is defined as [18]

$$F = \sum_1^N f_n \exp\left[i \frac{2\pi}{\lambda} (\overline{S} - \overline{S}_0) \cdot \overline{r}_n\right] = \sum_1^N f_n \exp[i 2\pi (hx_n + ky_n + lz_n)], \qquad (18)$$

where
  f_n is the atomic scattering factor of the nth of the N atoms in the unit cell,
  S_0 and S are vectors in the directions of the incident and the scattered x-rays, respectively,
  r_n is a vector from the origin of the unit cell to each of the atoms,
  hkl are the integer Miller indices of a set of lattice planes in the crystal, and
  x_n, y_n, z_n are the relative positions of atoms in the unit cell.

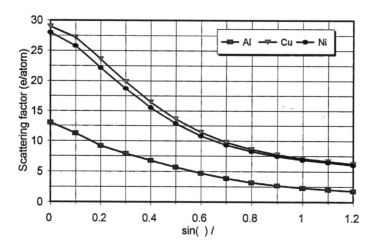

Figure 12. Atomic scattering factor for several materials plotted as a function of $(\sin\theta/\lambda)$.

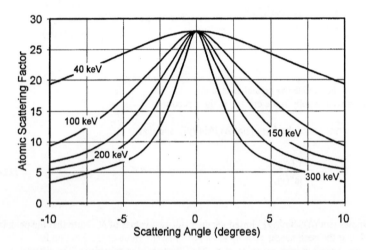

Figure 13. Atomic scattering factor at particular x-ray energies, for nickel, plotted as a function of scattering angle (2θ).

As an example, consider a nickel crystal, which has a face-centered cubic structure. The relative atomic positions in the unit cell are 000, ½½0, ½0½, 0½½. We use the second form of the structure factor in eq (18) to write

$$F = f_{Ni} [\, e^{\,i2\pi(h0+k0+l0)} \; + \; e^{\,i2\pi(h/2+k2+l0)} \; + \; e^{\,i2\pi(h2+k0+l/2)} \; + \; e^{\,i2\pi(h0+k/2+l/2)} \,] \; . \qquad (19)$$

If h, k, and l are mixed (even and odd), then $F = 0$. If the indices (hkl) are unmixed (all even or all odd), then $F = 4\, f_{Ni}$, and $FF^* = 16\, f_{Ni}^2$.

The intensity (I) scattered by a unit cell is then

$$I = I_0 \cdot F \cdot F^* \cdot \frac{e^4}{R^2 \, m^2 \, c^4} \; \frac{[1+\cos^2(2\theta)]}{2} \; (\frac{1}{4\pi \, \epsilon_0})^2 \; . \qquad (20)$$

Diffracted intensity will increase with the number of unit cells illuminated by the x-ray beam. For a small single crystal, illuminated by an unpolarized x-ray beam, the scattered intensity is [18]

17

$$I_p = I_0 \frac{e^4}{R^2\,m^2\,c^4} \frac{[1+\cos^2(2\theta)]}{2} \left(\frac{1}{4\pi\,\epsilon_0}\right)^2 F\,F^{\cdot} \prod_{i=1}^{3} \frac{\sin^2(\frac{\pi}{\lambda}(\overline{S}-\overline{S}_0)\cdot N_i\overline{a}_i)}{\sin^2(\frac{\pi}{\lambda}(\overline{S}-\overline{S}_0)\cdot\overline{a}_i)}, \quad (21)$$

where
  a_i are the vectors that define the crystal axes, and
  N_i is the number of unit cells in the crystal, in the direction a_i .

Note that the last terms are of the form $[\sin^2(Ny)/\sin^2(y)]$, and that

$$\frac{\sin^2(Ny)}{\sin^2(y)} \;\to\; \frac{(Ny)^2}{y^2} \;\to\; N^2 \quad as\ y \to n\,\pi \ . \tag{22}$$

Figure 14 shows $[\sin^2(Ny)/\sin^2(y)]$ plotted versus y, for two values of N. Note that the peak is highly localized, that the peak height goes up as N^2, and that the value of the function is approximately zero except at integer values of π. This illustrates the behavior of the diffraction process, which yields high diffracted intensities in certain directions and negligible intensities in all other directions. The product, $N_1^2\,N_2^2\,N_3^2$ will be very large even for a small crystal.

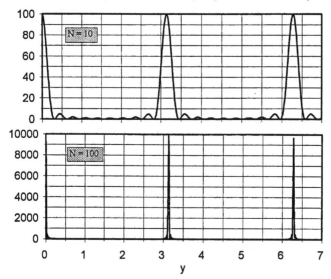

Figure 14. $[\sin^2(Ny)/\sin^2(y)]$ plotted versus y, for $N = 10$ and $N = 100$. There are maxima at $y=n\pi$.

18

Diffraction occurs when

$$\frac{\pi}{\lambda} (\overline{S} - \overline{S}_0) \cdot \overline{a}_i = n\pi .$$ (23)

These conditions, usually written

$$\begin{aligned}
(\overline{S} - \overline{S}_0) \cdot \overline{a}_1 &= h\pi \\
(\overline{S} - \overline{S}_0) \cdot \overline{a}_2 &= k\pi \\
(\overline{S} - \overline{S}_0) \cdot \overline{a}_3 &= l\pi ,
\end{aligned}$$ (24)

are known as the Laue equations. The integers hkl are the Miller indices of a particular crystalline lattice plane.

Often, a simpler form of the diffraction conditions is used. The Bragg equation

$$n\lambda = 1.24 \frac{n}{E} = 2d \sin(\theta) ,$$ (25)

where
  n is an integer,
  λ is the wavelength of the x-rays (nm)
  E is the energy of the x-rays (keV),
  d is the lattice spacing (nm), and
  2θ is the angle between the incident and diffracted x-ray,

expresses these conditions. Figure 15 illustrates that the Bragg equation describes the situation when the difference (AOB) in path length for x-rays "reflected" from adjacent lattice planes is an integer number of wavelengths (that is constructive interference occurs).

The diffraction interaction is thus a spatial (θ) and energy (E) filter. X-rays of higher energy coherently interfere at smaller angles. An x-ray imager placed in the diffraction field records the spatial location of diffracted x-rays. An energy-sensitive detector placed at the same location as the imager would record energy peaks associated with each of the diffraction spots.

Raising the temperature of the specimen decreases the diffracted intensity because the atoms move out of the mean position where diffraction occurs. Multiplying eq (21) by e^{-2M} (the Debye temperature factor) gives the correct intensity. The factor $2M = 16\pi\langle us \rangle^2 (\sin^2 \theta / \lambda^2)$, where $\langle us \rangle$ is the component of the mean displacement in the direction normal to the diffracting planes. This correction for temperature is applicable when all atoms in the small, diffracting crystal are the same.

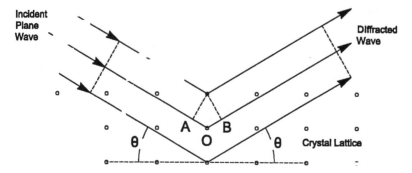

Figure 15. Diffraction occurs when the path length difference (AOB) between reflections is an integral number of wavelengths.

2.7 Implications of X-Ray Theory for High-Energy Transmission Diffraction

Probing the interior of a casting requires a transmission configuration. X-rays energies of over 100 keV are needed to penetrate the refractory oxide mold (5 to 10 mm wall thickness) and casting specimen (1 to 20 mm thick). Below, we explore the physics of using high-energy x-rays to produce transmission x-ray diffraction.

Consider the situation shown in Figure 16, of a beam of x-rays incident on a crystalline specimen contained within a casting mold. The primary x-ray beam is attenuated as it passes through the mold wall and a portion of the specimen, to a location where coherent scattering occurs. The scattered x-ray is attenuated along its exit path through the remaining specimen and exit mold wall. Attenuation losses (refer to Figs. 3 through 6) along entrance and exit paths are minimized by raising the energy substantially above that (5 to 20 keV) used in conventional x-ray diffraction systems to 150 to 300 keV.

The intensity of elastic scattering from a crystalline solid, given in eq (21), is the product of the squared structure factor (F) of the crystal, the Thomson scattering amplitude from an isolated electron, and the squares of the number of unit cells along the three crystalline axes. F is a function of the types of atoms in the crystal (through the atomic scattering factor, f), the configuration of the unit cell of the crystal, and the particular lattice planes which are involved in the scattering. For a face-centered cubic (fcc) crystal, such as nickel, the structure factor is equal to $4f$ if the Miller indices (hkl) are unmixed (all odd or all even), but is zero if the indices are mixed. The intensity of the scattering is the square of the amplitude. The structure factor is thus squared, to give $16f^2$ for an fcc material with lattice planes defined by unmixed Miller indices.

The dependence of the atomic scattering factor for nickel was plotted in Figure 13, as a function of the scattering angle for several x-ray energies. At low x-ray energies, the atomic

20

scattering factor is quite large, even for large scattering angles. However, with high x-ray energies, f is significant only for small scattering angles, that is, for a transmission geometry. For example, the scattering factor for nickel is 16.5 for 150 keV x-rays scattered at 3.5° from the direction of the incident x-ray beam (typical of our experimental geometry). The structure factor squared for this example is $16 \times (16.5)^2$, or 4356.

For a nickel crystal 0.001 mm × 0.001 mm × 0.001 mm, $N_1 = N_2 = N_3 = 2.8 \times 10^3$, so the product $N_1^2 \times N_2^2 \times N_1^2 \approx 5 \times 10^{20}$.

The atomic scattering factor in the forward direction, the enormous number of unit cells along the path of the primary beam (all with the same crystalline structure and orientation), in addition to the substantial intensity of high-energy x-rays which penetrate through a mold and specimen, all account for the efficiency of high-energy transmission diffraction.

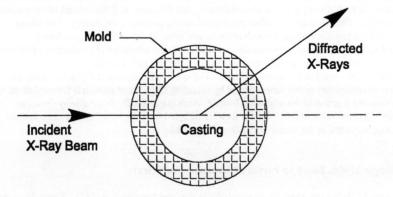

Figure 16. Geometry of transmission diffraction from a crystalline specimen surrounded by a casting mold.

3. MODELING TRANSMISSION DIFFRACTION

We studied the effects of properties and thicknesses of casting alloy and mold theoretically, to guide our development of transmission x-ray diffraction (XRD). A simple radiation transport model was developed to study x-ray diffraction from a specimen encased within a mold. The model has proven useful for determining the range of x-ray energies optimal for transmission x-ray diffraction. The model predicts the fraction of incident x-rays emerging as coherent radiation (which can produce diffraction), in the direction of a detector, from a metal specimen encased in a mold.

For in-situ casting measurements, x-rays from the source encounter a mold wall, then the specimen, another mold wall, and finally an x-ray detector. In addition, the beam may pass through some distance in air and perhaps furnace walls. The severity of the attenuation along this path depends on the x-ray attenuation coefficients, and densities, and thicknesses of the materials encountered. The attenuation coefficients depend on the incident x-ray energy. The model accounts for attenuation along the path (furnace and mold walls) of the entrance beam, attenuation in the specimen to the location of scattering, the efficiency of production of coherent scattering in the specimen, and the attenuation along the exit path (remaining specimen, mold wall, and furnace wall) of the scattered beam. The spatial distribution of diffracted radiation, which is approximated in this simple model by assuming that a fixed portion is forward-directed, determines the fraction of the scattering fluence which is potentially detectable by the x-ray detector. A fixed value of 0.007 for the forward directed fraction, based on our experimental geometry, was used in the model calculations reported here.

3.1 Simple Model, Based on Partial Attenuation Coefficients

Figure 17 shows the components necessary to perform transmission XRD. X-rays from the source are collimated into a narrow beam by a set of apertures. This narrow beam encounters an entrance mold wall, then the specimen, the exit mold wall, and finally an x-ray detector. X-rays entering the mold are attenuated. Attenuation also occurs in the specimen to the depth where coherent scattering occurs. X-rays diffracted from this volume element in the specimen are attenuated as they pass through the remaining specimen and the exit mold wall. The severity of the attenuation depends on the attenuation coefficients in the mold and specimen, the thicknesses, and densities of the materials, and the energy of the x-rays. Diffraction by an element of the specimen also depends on the specimen material and the x-ray energy. The processes of attenuation and diffraction are modeled to give insight into transmission XRD.

Let $I_0(E)$ be the x-ray intensity (at energy E) incident upon the detector with no specimen or mold present. The x-ray intensity which is transmitted through the entrance mold wall is

$$I_1(E) = I_0(E) \ e^{(-\mu_{wi}(E) \ \rho_w \ W_1)} \ , \tag{26}$$

where

μ_{wt} (E) is the total mass attenuation coefficient of the mold wall,
ρ_w is the density of the mold wall, and
W_1 is the thickness of the entrance mold wall.

Let I_2 (E) be the intensity of coherent scatter which emerges from the specimen. I_2 depends upon: (1) the attenuation in the specimen from the entrance surface to the location where coherent scattering occurs, (2) the efficiency of generation of coherent scatter, and (3) the attenuation of the coherent radiation as it passes through the remainder of the specimen. Summing the effects of the loss and generation mechanisms for all elements through the thickness of the specimen gives

$$I_2(E) = I_1(E) \sum_{n=1}^{N} e^{-\mu_{st}(E)\rho_s n\, dx} \left[K\left(1-e^{-\mu_{sR}(E)\rho_s dx}\right)\right] e^{-\mu_{st}\rho_s (L-n\, dx)} , \qquad (27)$$

where

μ_{st} (E) is the total mass attenuation coefficient in the specimen,
ρ_s is the density of the specimen,
N is a large number which partitions the specimen length L into length segments dx,
K is the fraction of coherently scattered photons which potentially could contribute to a transmission diffraction pattern, and
μ_{sR} (E) is the partial mass attenuation coefficient for coherent scattering in the specimen.

The negative exponential in the second term of the summation represents the loss from the narrow beam due to production of coherent radiation. Since the scattering is isotropic, a fraction

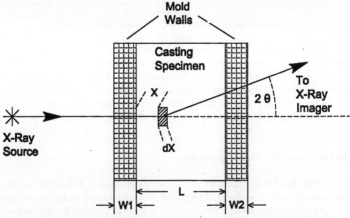

Figure 17. Geometry used in developing an analytical model for transmission XRD by a casting specimen encased by mold wall.

23

(K) of the coherent intensity is forward directed. It is this forward-directed scattering which will produce the transmission XRD pattern. The entire second term in the summation thus gives the fractional generation of coherent scattered radiation which emerges from a point within the specimen in the forward direction. The value of K is determined by the geometry of the diffraction experiment. Since the pattern is observable only at the x-ray imager, we assume K is the fractional solid angle subtended by the imager from the point of scattering.

Equation (2) can be symbolically simplified to

$$I_2(E) = I_1(E) \; K \; N \; [e^{-\mu_{st}(E)\rho_s L} - e^{[-\rho_s(\mu_s f(E)L + \mu_{sR}(E) dx)]}] \; . \tag{28}$$

The intensity of the coherent x-rays which emerge from the exit mold wall is given by

$$I_3(E) = I_2(E) \; e^{(-\mu_{wt}(E)\rho_w W_2)} \; , \tag{29}$$

where
W_2 is the thickness of the exit mold wall.

Combining eqs (26), (28), and (29) gives an expression for the intensity of coherent x-rays emerging from the exit mold wall in terms of the thicknesses, the densities, and the x-ray attenuation coefficients of the mold walls and specimen. It is convenient to compute the ratio $I_3(E)/I_0(E)$ which is the fraction of the incident x-rays which emerge in the forward direction, from the mold-encased specimen, as coherent radiation.

3.1.1 Advantages and Limitations of the Attenuation-Coefficient Model

The simple model for transmission diffraction gives the correct energy dependence for small scattering angles and can be used to compare diffraction efficiencies for variations of specimen and mold thicknesses and densities. However, because the model does not account for the constructive interference of coherent scatter from atoms in a crystalline solid, it cannot predict intensities for reflections from particular lattice planes. For the same reason, the absolute efficiencies are incorrect, and K is actually a function of the diffraction angle and energy.

3.1.2 Model Predictions for Metal Specimens

This simple model has proven useful for investigating transmission diffraction for the materials we used in our experiments. The effects of specimen thickness, and of mold material surrounding the specimen were studied extensively. Partial and total x-ray attenuation coefficients were

24

computed using the XCOM software package [22]. The value of K was chosen as 0.007, being representative of our laboratory XRD apparatus (imager 50 mm in diameter located 150 mm from the diffracting specimen).

Figure 18 plots $I_3(E)/I_0(E)$ for a specimen of pure aluminum. The efficiency of transmission XRD is a strong function of both the specimen thickness and the incident x-ray energy. The peaks in the plots indicate an optimal range (50 to 100 keV) of x-ray energies for transmission XRD on these aluminum specimens. The optimum energy yields the highest intensity of diffracted x-rays which are able to penetrate through the specimen.

Remember that plots such as shown in Figure 18 include the effects of both attenuation and diffraction. For a thin specimen, there is very little attenuation. However, the irradiated volume of a thin specimen does not generate much diffraction. A thicker specimen attenuates more of the incident and diffracted x-rays, but produces a stronger diffraction pattern. For high-energy x-rays and the range of aluminum thicknesses shown, diffraction dominates attenuation.

Figure 19 shows plots of model-generated efficiency of diffraction for a nickel specimen. Because the mass attenuation coefficients and density of nickel are higher than those for aluminum, the efficiency of transmission diffraction is lower. For specimens 2 to 10 mm thick, diffraction again overwhelms attenuation. But for specimens of greater thickness (20 to 25 mm), attenuation of the incident and diffracted x-rays begins to dominate and the efficiency of diffraction decreases. Note also that the optimal x-ray energy for transmission diffraction is higher when a nickel specimen is probed than that for an aluminum specimen of identical thickness.

3.1.3 Model Predictions for a Mold-Encased Specimen

Figure 20 shows plots of effiency of diffraction for a nickel specimen 10 mm thick sandwiched between two pieces of mold material, each of the thickness indicated in the legend. Attenuation coefficients for the mold material (a mixture of alumina, silica, and zirconia) were computed using the XCOM software. The density of a sample of the material was 2.6×10^3 kg/m^3. For x-ray energies over 150 keV, even mold-wall thickness as large as 7.5 mm decrease the XRD intensity only by about one-half.

3.1.4 Modeling Various Fractions of Liquid and Solid

Anticipating that we would use the XRD technique to probe the "mushy" zone (area of dendritic solidification containing both solid and molten material) of a casting, we modeled a specimen volume containing both liquid and solid. The liquid portion of the specimen was assumed to attenuate the incident and diffracted x-ray beams. The correct (lower) density of the liquid was used. The solid portion of the specimen attenuated both the incident and scattered x-rays as well and produced coherent scatter. A 6 mm thick section of casting was modeled, with various fractions of the thickness being comprised of molten metal and crystalline solid. Energies

25

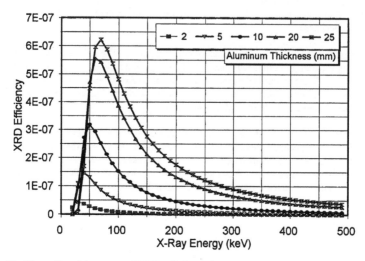

Figure 18. Plots of model-generated XRD efficiency from an aluminum specimen. The legend gives the thickness, in the range 2 to 25 mm.

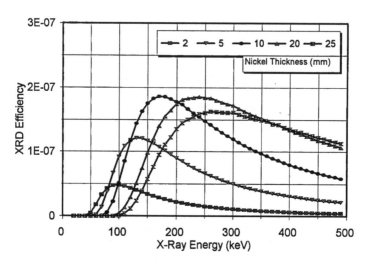

Figure 19. Plots of model-generated XRD efficiency from a nickel specimen. The legend gives the thicknesses, in the range 2 to 25 mm.

26

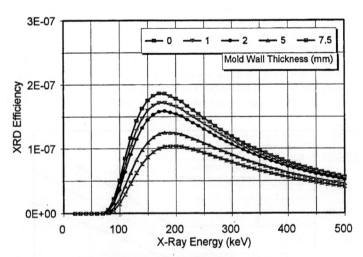

Figure 20. Plots of model-generated XRD efficiency from a nickel specimen 10 mm thick surrounded by mold material.

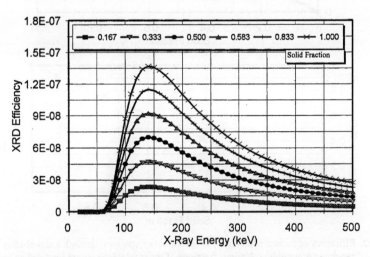

Figure 21. Modeled efficiency of transmission XRD through a nickel-alloy specimen 6 mm thick. Various fractions of the total thickness are modeled as liquid and crystalline solid.

in the range from 20 to 500 keV were considered. Figure 21 shows plots of the XRD efficiency (fraction of the incident x-ray intensity which emerges as coherent radiation) as a function of the x-ray energy. For low fractions of solid the efficiency of diffraction and the intensity of diffraction spots are low. Efficiency and intensity increase as more of the specimen thickness is in a physical form (crystal) which diffracts the incident x-ray beam. Figure 22a plots values of efficiency (from Fig. 21) at several x-ray energies. Figure 22b plots these efficiencies, normalized to the efficiency when the entire specimen is a single diffracting crystal (fraction of solid is 1). The relationship between efficiency and fraction of solid departs only slightly from a linear one. The slightly concave shape of the curve results from differences between the densities of the liquid and solid.

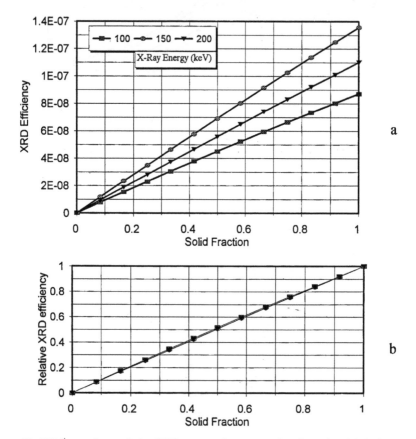

Figure 22. Efficiency of transmission XRD, at several x-ray energies, through a nickel-alloy specimen 6 mm thick. Various fractions of the total thickness are modeled as liquid and crystalline solid. The plots in (a) are normalized in (b).

3.2 Fundamental Physics Model

We showed in section 2 that for a small single crystal, illuminated by an unpolarized x-ray beam, the scattered intensity could be calculated. This intensity is not measureable, but the total diffracted energy (integrated intensity) is. The total diffracted energy is given by

$$E = \frac{I_0}{\omega} \frac{e^4}{R^2 m^2 c^4} \frac{[1+\cos^2(2\theta)]}{2\sin(2\theta)} \left(\frac{1}{4\pi \epsilon_0}\right)^2 FF^* e^{-2M} \lambda^2 \frac{V}{v_a^2}, \tag{30}$$

where
ω is an angular rate of rotation of the specimen, required to insure that all area under the interference functions, eq (22), has been integrated,
V is the volume of a small segment of the specimen illuminated by the incident x-ray beam,
v_a is the volume of the unit cell of the crystal.

The coherent-scattering term in eq (27) can be replaced by eq (30). Attenuation in the mold and casting, given in eqs (26), (27), and (29) remains the same. The energy-dependence of diffraction comes into eq (30) through the energy-dependent atomic scattering factors which are included in the structure-factor term FF^*. The Lorentz-polarization factor term is also a function of x-ray energy, through the scattering angle 2θ.

Figure 23 shows the efficiency of transmission XRD modeled by eqs (26), (27-modified), (29), and (30), for the (111) reflection from nickel. The energy dependence is similar to that shown in Figure 19. However, the absolute intensity is correct.

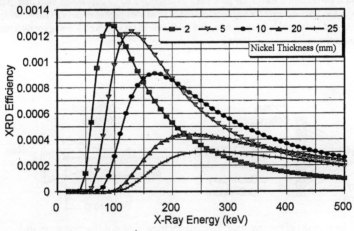

Figure 23. Plots of model-generated XRD efficiency from a nickel specimen.

29

4. APPARATUS

4.1 Equipment for High-Energy, Transmission Diffraction

Figure 24 illustrates the key elements in the apparatus used for transmission XRD measurements. The x-ray tube produces a high-intensity, polychromatic source of x-rays restricted by collimators to a small circular beam. A fraction of the x-rays interacting with the sample is diffracted. An x-ray imager detects the x-ray pattern incident on its two-dimensionally sensitive area. An energy-sensitive detector may instead be used, in place of the imager, to measure the spectra of the primary and diffracted x-rays.

The components of the diffraction system are described first, followed by descriptions of the laboratory apparatus used for room temperature studies, and the furnaces used for melting and solidificaton experiments. Three separate furnaces were assembled for our tests—a low temperature test cell for melting gallium, a medium temperature gradient furnace for melting aluminum and copper, and a high-temperature directional solidification furnace (for growing single-crystal castings of nickel-alloy).

4.1.1 X-Ray Source

Two sources of x-rays were used in our experiments. Both were metal-ceramic, tungsten-anode tubes with a beryllium exit window, intended for use in industrial radiographic imaging. A 160 kV tube head (Fig. 25) was used in early experiments. However, modeling indicated the need for x-rays of higher energy when a turbine-blade casting was to be probed. A new x-ray tube (Fig. 26) and high-tension transformer permitted experiments to be performed with tube potentials as high as 320 kV. An increased x-ray tube potential produces higher intensity characteristic radiation lines and bremstrahlung spectrum, and adds new x-rays of higher energy which more easily penetrate the mold-encased casting.

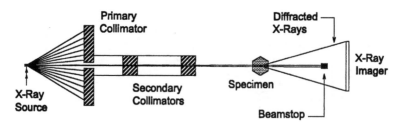

Figure 24. Components of the real-time, high-energy transmission diffraction apparatus.

Figure 25. End-window x-ray tube head, 160 kV maximum potential. A short collimator is attached to the tubehead.

Figure 26. X-ray tube head, 320 kV maximum potential. The collimator is shown affixed to the tubehead in this photograph.

The choice of x-ray source diameter (focal spot size) of 1.2 mm or 4 mm (320 kV tubehead), or 0.2 or 3 mm (160 kV tubehead) is selected by activating one of the two filaments in the tube. The small (1.2 mm) focal spot size of the 320 kV tubehead proved most useful in our experiments because its size corresponds to the diameter (1 mm) of the collimating apertures. X-ray intensity from this source was significantly higher at the specimen, by a factor of 10, than that obtainable from the 0.2 mm source in the 160 kV tube head. The higher intensity yielded a higher signal-to-noise ratio in the XRD images.

Controls on the constant-potential x-ray generator permit x-ray tube voltages to be varied over the 15 kV to 320 kV range. A tube current of 3 mA is possible at 320 kV. For some experiments a means for reducing the x-ray source intensity was required so the diffraction signals did not overwhelm the detector system. A lower intensity was obtained by modifying the controller. Tube currents as low as 0.05 mA with incremental changes of 0.05 mA were possible.

A pinhole-camera image of the x-ray source (focal spot) was be made by placing a small (0.5 mm diameter) aperture in a lead plate between the x-ray tube and an imaging device. The dimensions of the source and its uniformity can be assessed from the image. Figure 27 shows an image produced by the 1.2 mm diameter focal spot of the 320 kV tubehead.

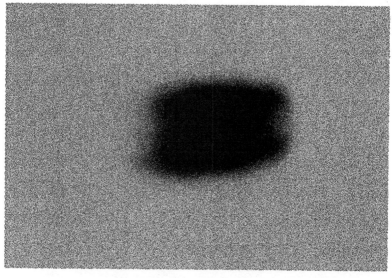

Figure 27. Pinhole camera image of the 1.2 mm diameter focal spot of the 320 kV tubehead.
X-ray parameters were 320 kV, 1 mA.

4.1.2 Leakage Radiation and Supplemental Shielding

During experiments with a high-sensitivity x-ray detector, the presence of undiffracted radiation was evident. An investigation was made to determine the source of the unwanted radiation. An x-ray pinhole camera was constructed (with lead walls 10 mm thick and a pinhole 2 mm diameter) to image the location of the radiation leaking from the x-ray tube housing. An electronic x-ray imaging device was used to provide a real-time image of any undesirable radiation sources within the field of view of the pinhole camera. X-rays were found to be leaking from the area surrounding the collimator (Fig. 28). Supplemental x-ray tube shielding (lead, 9.5 mm thick), alleviated the leakage radiation problem and improved the signal-to-noise ratio of the diffraction images.

4.1.3 Collimators

A beam-restricting collimator (Fig. 29) was designed and fabricated. A triple-aperture design (Fig. 1) was employed to minimize beam divergence and optimize x-ray intensity [24]. Interchangeable sets of 9 mm thick lead disks can be used to select beam diameters of 0.3 mm, 1 mm, 1.5 mm, or 2 mm. An alignment laser, placed in an opening in the base of the collimator, projects a visible beam along the same path as that of the x-ray beam. This permitted fast, safe, and simple alignment of the beam-stop and specimen.

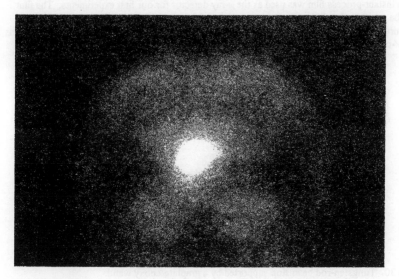

Figure 28. X-ray pinhole camera image of radiation leakage from the 320 kV tubehead.

33

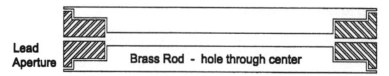

Figure 29. Drawing of the collimator which restricts the x-ray beam to 1 mm diameter.

4.1.4 Beamstop

The primary beam transmitted through the specimen is often of high intensity. Blocking this beam from the detector improves the contrast of images and prevents an energy-sensitive detector from being overwhelmed by high count rates. A small tungsten rod (4 mm diameter and 6 mm thick) was positioned in the center of the primary x-ray beam emerging from the specimen, to act as a beamstop. X-rays diffracted from the sample pass to the sides of the beamstop and are imaged, while the undiffracted primary beam is severely attenuated. The tungsten disk was suspended by a graphite/epoxy composite support (Fig. 30). The low attenuation in the composite minimizes the shadow it casts on the x-ray images.

4.1.5 X-Ray Imager

An instant-process film was used as the x-ray detector for our first experiments. The film could be positioned to intercept various portions of the x-ray scattering from the sample. Exposure times of less than one minute were sufficient to record discernable diffraction patterns when x-ray source parameters of 75 kV and 18.5 mA were used.

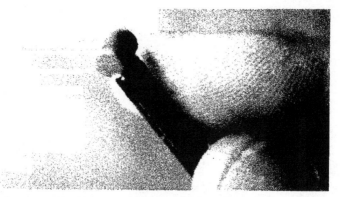

Figure 30. Tungsten-rod beamstop supported by a graphite/epoxy wand.

34

A real-time x-ray imager, designed specifically for x-ray diffraction, was acquired and used for all experiments reported in this paper. This unit (Fig. 31) uses a scintillator screen, coupled to an image intensifier, which is then coupled to a charge-coupled device (CCD) camera sensitive to low light levels. Fiber-optic coupling between the scintillator, image intensifier, and CCD camera transfers light extremely efficiently. Image intensifier and video gain were adjustable. The imager was originally equipped with a 70 μm thick gadolinium oxysulfide (Gd_2O_2S) scintillator, resulting in a system with a resolution of approximately 80 μm. The field of view of the imager is approximately 40 mm wide by 30 mm high. Imager specifications are summarized in Table 1.

Table 1. Specifications for the real-time x-ray crystallography imager.

Scintillator	70 μm thick polycrystalline gadolinium oxysulfide
Input image size	50 mm diagonal, 3:4 aspect ratio
Optical coupling	Fiber optic throughout
Video format	CCIR 625 lines at 50 Hz
Video output	1 V peak-to-peak into 75 Ω
Gain range	1 to 1000, user adjustable
Physical size	287 mm L, 80 mm H, 107 mm W

Figure 31. Real-time x-ray imaging device. The imager comprises a scintillator, image intensifier, and a CCD camera.

We used a calibration grid (5 mm line spacing) fabricated from thin copper-coated printed-circuit-board material to measure the spatial uniformity of the imager. The grid was affixed to the front of the scintillator and placed in a low-intensity x-ray field. Figure 32 shows the resultant image. Note the pincushion distortion of the grid. The spatial distortion in a recorded diffraction image can be corrected by reference to the calibration image.

To improve the signal-to-noise ratio of the XRD images, we examined the efficiency of the x-ray imager. To be detected, diffracted x-rays must be absorbed in the scintillating layer at the front of the imager and produce light. The scintillator material and its thickness determine what fraction of the incident x-ray flux is absorbed. There is a tradeoff between good spatial resolution with a thin layer and high absorption in a thick layer. The x-ray imager (with the gadolinium oxysulfide scintillator) was optimized for high spatial resolution at low to moderate energies. As we pushed the x-ray energy in our experiments ever higher for increased transmission through furnace components, mold, and specimen, the efficiency of the imager decreased. Although acceptable transmission XRD images could be obtained, an improvement was possible by replacing the gadolinium oxysulfide scintillator with one having a higher attenuation at energies up to 320 keV. A 6 mm thick fiber-optic glass scintillator [27], which had been acquired for work at 160 keV, was affixed to our imager. The fiber-optic scintillator produces extremely low lateral spread of light and high-efficiency transfer of light to the image intensifier. Figure 33 shows the absorption calculated for the scintillators. The glass scintillator absorbs a much greater number of x-ray photons than the gadolinium oxysulfide, particularly at high energies.

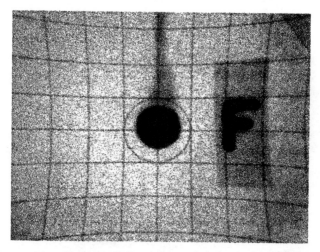

Figure 32. Image of the calibration grid (5 mm line spacing) showing pincushion distortion inherent in the x-ray imager.

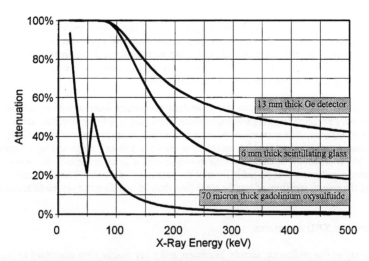

Figure 33. Absorption of x-rays in the 70 µm thick gadolinium oxysulfide scintillator and the 6 mm thick fiber optic scintillator. For comparison, the absorption in our energy-sensitive germanium detector is shown. The discontinuity at 50.2 keV in the curve for gadolinium oxysulfide occurs at the K absorption edge of gadolinium.

Acquisition and storage of the video frames from the PAL-format imager is performed with an eight-bit, monochrome frame grabber installed in a personal computer. A coprocessor board, connected to the frame grabber, speeded image manipulations and frame averaging (to increase signal-to-noise ratio). Frame averaging could be performed at a rate of 25 frames per second. Image acquisition and processing software provided the ability to perform dynamic range expansion and compression, image averaging, and image subtraction, as well as image filtering to enhance the acquired images.

A multiformat video cassette recorder was used to record radiographic and diffraction images during all experiments. This served as a backup for direct image acquisition with the frame grabber, and was useful in cases where the image was rapidly changing (dynamic events such as mold filling). Video frames could be digitized after the experiments by replaying the video tape through the frame grabber. The image degraded almost imperceptibly compared with the image acquired directly from the x-ray imager.

4.1.6 Energy-Sensitive Detector System

In addition to x-ray diffraction images, we have investigated the use of energy-sensitive detectors. The diffraction interaction is an energy as well as spatial filter. An x-ray imager placed in the diffraction field records the spatial location of diffracted x-rays. An energy-sensitive

37

detector placed at the same location as the imager would record energy peaks associated with each of the diffraction spots. To explore the information contained in the energy spectrum of diffracted x-rays, we began to use an energy-sensitive x-ray detector. Preliminary experiments using a large (75 × 75 mm) NaI scintillator appeared promising; however, the poor energy resolution precluded observing the expected spectral peaks.

An improved system (Fig. 34) includes an intrinsic germanium detector with an integral field-effect-transistor preamplifier, a high-voltage detector bias supply, a linear amplifier, and a multichannel analyzer. The germanium detector was a thick, large-area planar type. Components of the system were optimized for high sensitivity in the 40 to 500 keV range and high counting rate (2×10^5/s) capability. Specifications for this equipment is given in Table 2. For some experiments, a collimator was placed in front of the germanium detector. This permitted examining the energy spectrum of smaller areas (less than 36 mm diameter) in the diffraction field.

4.2 Laboratory XRD Apparatus

The x-ray source collimator, sample, beamstop, and x-ray imager were assembled as shown in Figure 35. The specimen was placed in the collimated x-ray beam using a versatile fixture. Both the imager and the furnace were on separate rotating stages. The imager stage could be manipulated under computer control. The furnace could be raised and lowered remotely, as well. The furnace and imager were also attached to roller-bearing stages so that the x-ray source-to-sample distance could be varied, and the furnace could be moved side-to-side in the x-ray beam. The fixture rigidly supported the x-ray source, sample and x-ray imager, and had provisions for quantitatively determining positions and angulations.

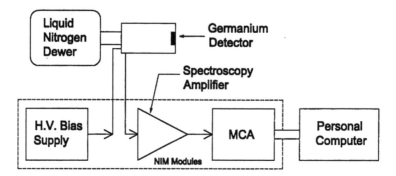

Figure 34. Energy-sensitive detector system. The components were optimized for extreme counting-rates, so high-intensity areas in the x-ray field could be evaluated.

Table 2. Specifications for the energy-sensitive x-ray detector system.

Detector type	Intrinsic germanium, planar geometry
Detector dimensions	36 mm diameter × 13 mm thick
Detector bias	1500 Volts, negative polarity
Detector energy resolution	0.332 keV at 5.9 keV, 6 µs shaping time, 0.573 keV at 122 keV, 6 µs shaping time, 0.754 keV at 122 keV, 1 µs shaping time, and count rate of 10^5 s^{-1}
Amplifier	Linear spectroscopy amplifier, 2.5 to 1500 gain, 0.5 to 10 µs shaping times, pulse pile-up rejector, automatic pole-zero adjustment, baseline restorer
Multichannel analyzer	16384 channels, $(2^{31}-1)$ counts per channel, counting rate throughput $>10^5$ s^{-1}, PC-based interface for control and data transfer

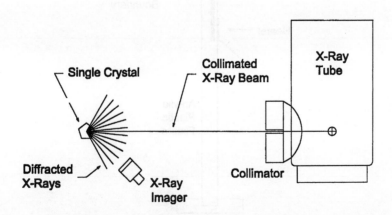

Figure 35. Drawing of the laboratory apparatus used for XRD experiments at room temperature.

4.3 Gallium Test Cell

It was convenient to begin our diffraction experiments with gallium. Its low (28 °C) melting point permitted studies to be made close to room temperature, and the large (3 percent) change in density between the solid and liquid was great enough to produce a discernable difference in brightness on a radiographic image. The ability to independently determine the position of the liquid-solid boundary (brighter versus darker areas of the radiograph) provided a means for validating the spatial performance of the XRD sensor.

A container (Fig. 36) was fabricated from polymethylmethacrylate. A temperature gradient could be established between the 0 °C temperature (ice) at the top of the cell and an electrically heated region lower in the cell. A temperature controller connected to the heater produced a steady-state boundary between solid and liquid gallium. The lower-density solid is at the top of the test cell. The position of the probing x-ray beam could be moved into the solid or liquid by a remotely controlled positioning fixture. We could switch between the XRD sensor and radiographic imager without changing position by removing the collimating apertures in the x-ray beam.

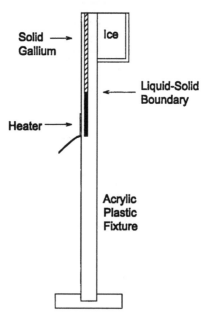

Figure 36. Gallium test cell. A steady-state boundary between solid and liquid could be established by controlling the heating element.

4.4 Medium-Temperature, Gradient Furnace

Since the aim of developing the x-ray diffraction technology was to study metal solidification, a furnace was required to heat and melt the specimen. Figure 37 shows the controlled-gradient furnace. It is a vertical tube furnace with separate controls for both the upper and lower regions of the tube. By adjusting the temperatures of the two zones, a temperature gradient could be established for solidification in the region of the furnace probed by a collimated x-ray beam. The specimen to be melted was placed in a closed-end quartz tube (25 mm i.d., 2.2 mm wall thickness, 300 mm length). The type K thermocouples used for temperature measurement and control were protected by closed-end quartz tubes 1 mm i.d., 28 mm long. The furnace was designed to reach 1100 °C. It has been tested to 1100 °C and has proven suitable for controlled melting and solidification of aluminum and copper. We have repeatedly cycled through the temperature range within which the metal sample melts and solidifies.

The solidification zone of the gradient furnace is established by adjusting the temperature of the upper heater to be above the metal-melting point and that of the lower heater below the melting point of the specimen. Normally, the liquid-solid boundary is positioned within the area occupied by the ceramic foam spacer. The foam spacer provides a path of low attenuation for the probing x-ray beam.

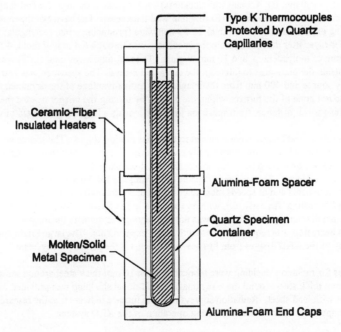

Figure 37. Gradient furnace for melting and solidification studies of aluminum and copper.

41

In our melting experiments on aluminum and copper, the metal was enclosed in a quartz tube. At the elevated temperatures in the gradient furnace (700 °C for aluminum and 1100 °C for copper) the metal reacted with the tube, scavenging elements from it. Additionally, because of the difference in thermal contraction of the metal and quartz, the tube often cracked upon cooling. To alleviate these two problems, the specimen receptacle was changed to graphite, which was less reactive with the metals and more tolerant of thermal cycling. However, the graphite crucible cannot be used in an oxygen environment because of the possibility of oxidation and fire. Therefore, a glass enclosure was designed and built so the gradient furnace could be flooded with inert gas. The additional borosilicate glass of the enclosure walls (3 mm thick) did not appreciably alter the x-ray diffraction images obtained.

4.5 Directional-Solidification Furnace

A resistively heated directional-solidification (DS) furnace (Fig. 38) was acquired. The vacuum furnace (100 mm bore) was capable of producing 1700 °C temperatures. The x-ray source (A) and collimator (B) are on the far right. The small-diameter x-ray beam passes through a borosilicate glass port 10 mm thick into the furnace. The beam then encounters molybdenum resistance-heater windings (D) of 1.6 mm diameter. The wire is wound on an aluminum oxide support tube (E) (100 mm ID, 4.5 mm wall thickness). After passing through the coil support, the x-ray beam enters the casting mold (F) (6.4 mm wall thickness). The cavity of the mold for the first melting experiment was 3 mm thick by 38 mm wide (producing a thin rectangular bar). The exit path x-rays, after diffraction from the specimen, was through 6.4 mm of mold, 4.5 mm of alumina, 1.6 mm of molybdenum, and 10 mm of glass. The real-time x-ray imager (G) was positioned outside the glass port to intercept the diffracted x-rays. The specimen was 780 mm from the x-ray source and 400 mm from the imager. We took advantage of the asymmetry in the location of the hot zone of the furnace within the bell jar, by placing the imager nearest the specimen. This shorter distance, from specimen to imager, yields a larger angular field of view.

The x-ray source and imager were attached to two-axis motion stages. The source and imager could be scanned in unison (horizontally and/or vertically) to probe different regions of the specimen. However, by leaving the x-ray source fixed and scanning the imager, a large virtual field of view was achieved. The system is capable of moving the 130 kg x-ray head in submillimeter increments and at speeds more than sufficient for following the liquid-solid boundary in a DS casting. An early test, with no specimen in the furnace, disclosed that the primary x-ray beam intensity at the imager was not high enough to damage the imager. Therefore, no beamstop was required in these diffraction experiments. The large bright spot visible in many of the XRD images from furnace experiments is the primary x-ray beam.

Enclosures for radiation shielding were fabricated, from 16 mm thick steel around the x-ray tube and 9.5 mm thick steel around the x-ray imager. Additional shielding was provided by lining the enclosures with lead sheet. Radiation surveys near the furnace indicate that the researchers can safely occupy areas near the furnace during operation of the XRD system.

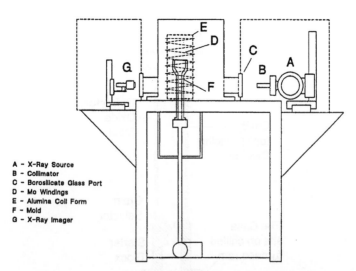

A - X-Ray Source
B - Collimator
C - Borosilicate Glass Port
D - Mo Windings
E - Alumina Coil Form
F - Mold
G - X-Ray Imager

Figure 38. Configuration of the directional-solidification furnace fitted with the XRD equipment. The dashed lines indicate lead-lined steel enclosures used for radiation shielding.

In normal operation, a mold containing a metal charge in a crucible at its top is placed in the load chamber of the furnace. After a 10 μm vaccum has been established the gate valve is opened and the vertical ram raises the mold into the hot zone of the furnace. The alloy charge melts, filling the mold. Solidification initiates at the base of the mold where the alloy contacts the water-cooled ram. The mold is slowly (150 mm/hr) withdrawn through the steep temperature gradient established between the hot zone and a chill plate beneath it. A single-crystal casting is produced.

Figure 39 shows a mold typical of that used in the single-crystal casting experiments. The mold is fabricated by repeatedly coating a wax replica of the part to be cast with a slurry of refractory oxides (alumina, silica, and zirconia). The slurry used for initial coats is very fine, to produce a smooth inner coating. The slurry for outer coats is increasingly coarse. The outer surface of our molds had an rms roughness of about 2 mm. Roughness varied from mold to mold. Once the requisite mold thickness has been achieved, the wax replica is removed by heating the mold. The mold is then baked at high temperature to harden it. To study different effects, we used molds with circular, triangular, or rectangular cross sections (Fig. 40).

During casting experiments, a metal alloy charge is placed in the crucible. When the mold is raised into the hot zone of the furnace, the alloy melts and flows into the mold. The alloy solidifies as polycrystals (grains) in the starter block. The fastest-growing grain reaches the grain selector (a corkscrew shaped section of the mold) first and blocks growth of all other crystals with differing orientation. If the thermal conditions are correct, growth of a single-crystal continues upward in the mold as the mold is slowly withdrawn through the furnace.

43

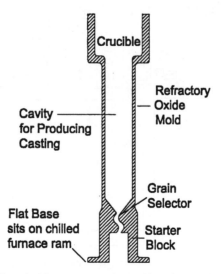

Figure 39. Side view of a typical investment casting mold used for producing single-crystal castings.

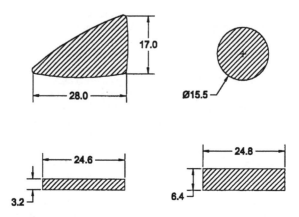

Figure 40. Cross sections of the four types of molds used in our casting experiments. All dimensions are in mm.

44

4.5.1 Modifications to the DS Furnace

The geometry of x-ray source, specimen, and x-ray imager for the tests in the directional solidification furnace was very different from that of the moderate-temperature XRD apparatus. The diffracted x-ray intensity arriving at the x-ray imager is much lower because the distance between the x-ray source and imager is much larger, and there is a great deal of attenuation from the many additional materials encountered along the x-ray path through the furnace. Figure 41 shows a cross section of the furnace as it was configured when received.

Initial experiments with the DS furnace indicated that the region of solidification in the casting was beneath the field of view of the XRD sensor. The hot zone of the furnace had to be turned off after melting occurred, to drive the solidification upward into the region probed by the sensor. To correct this problem, we rearranged components in the furnace and added new elements.

The hot zone (alumina core with molybdenum windings) of the furnace sat directly on the chill plate. Solidification occurred at the location of the chill plate or below it. This area is inaccessible for XRD because of the locations of the diffusion pump inlet, the chill plate, and its cooling coils. A 38 mm thick piece of alumina foam was used as a spacer between the chill plate and the hot zone. The foam has a structural rigidity sufficient to support the hot zone, but a density (240 kg·m^{-3}) low enough to provide good x-ray transmission. At the same time the foam spacer was added, the chill coils were raised slightly. This was done to preserve the high temperature gradient in the furnace which drives solidification in the vertical direction.

Extension of the top of the metal bell jar was required to raise the hot zone. Figure 42 shows the section added to the bell jar. Cooling coils were soldered to the extension to prevent overheating.

We have made further modifications to the directional solidification furnace. These changes were made to shift the solidification zone in the casting to coincide with the field of view of our XRD sensor (x-ray source and imager) and to enlarge the region of the casting we could

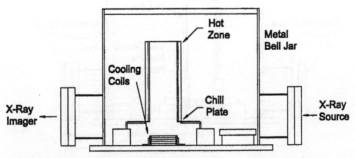

Figure 41. Side view of the bell jar in the directional solidification furnace as it was configured when received.

45

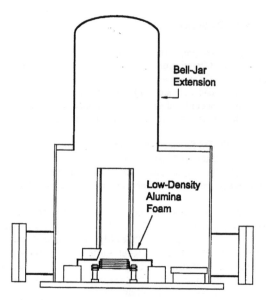

Figure 42. Side view of the bell jar in the directional solidification furnace after recent modifications.

investigate. Figure 43 shows a cross section of the furnace after the latest modifications. The height of the alumina foam spacer was increased from 38 mm (previous modification, Fig. 42) to 60 mm, thus enlarging the vertical range we could probe with the x-ray sensor.

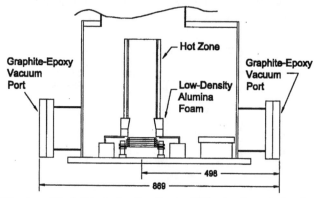

Figure 43. Side view of the bell jar in the directional solidification furnace after the final modifications to enlarge the field of view for XRD.

The ceramic foam spacer, beneath the hot zone, provides a path of low x-ray attenuation, free of interfering structures. The chilling coils were lowered slightly below the chilled plate. New support feet (fabricated from alumina foam) were constructed, to drop the top of the chilled plate to the height of the diffusion-pump inlet.

The added height of the hot zone of the furnace above the baseplate necessitated construction of a spacer (with water-cooled passages) for the vertical ram. The casting mold was then able to be positioned as before, vertically centered within the hot zone of the furnace.

The alignment of the crystalline planes in the casting can vary rotationally about the axis of the casting and can deviate from the vertical axis. A means for orienting the probing x-ray beam to obtain strong diffraction spots was devised. The vertical ram of the furnace was altered to permit rotation of the mold (and specimen). The rotation allows us to position the solidifying metal such that Bragg conditions are satisfied and diffraction spots can be observed. After the specimen rotation capability was added to the furnace (in August 1996) strong diffraction patterns from the crystalline alloy castings could be obtained during every experiment.

4.5.1.1 Modeling of Transmission XRD From a Casting in a Directional-Solidification Furnace

Our XRD model was used to predict the efficiency of the transmission XRD process for studying solidification of a specimen within the high-temperature directional-solidification furnace. Modeling (Fig. 44) shows the advantage which can be achieved if the hot zone components (alumina core and molybdenum windings) can be removed from the path of the XRD sensor. An additional increase in diffracted intensity can be expected if the borosilicate glass ports of the furnace are replaced with low attenuation graphite epoxy ports.

We had ports 9.5 mm thick fabricated from a graphite epoxy composite (quasiisotropic layup). The high modulus of the composite makes it ideal for a vacuum port, and the very low atomic numbers and densities of the fiber and matrix exhibit extremely low x-ray attenuation. Less of the incident and diffracted x-ray beams are attenuated in the ports, yielding a higher diffraction spot brightness. X-ray transmission was measured to assess the improvement in transmission due to use of the new graphite/epoxy ports. The two new ports attenuated the beam by only 15 percent, as opposed to the 44 percent attenuation by the glass ports (an improvement of a factor of 1.5).

4.6 Radiation Safety

Radiation shielding for our x-ray diffraction facilities was considered before begining any of the experiments. The reflection XRD, laboratory XRD, gallium test cell, and the medium-temperature gradient furnace studies were performed in a lead-shielded room. The experiments with the directional solidification furnace required adding radiation shielding around the source, the imager, and along the beam path through the furnace. An enclosure made of steel plate for the x-ray tube (16 mm thick) and for the x-ray imager (9.5 mm thick) provided adequate shielding. A researcher could safely occupy the area near the furnace during XRD tests (Fig. 45 and Table 3). The low radiation dosages are largely a result of the extremely small (1 mm diameter) x-ray beam used to probe the casting.

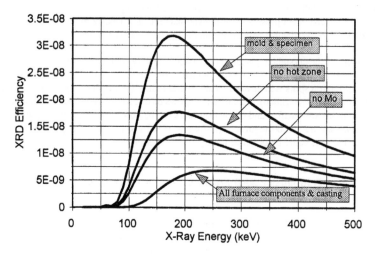

Figure 44. Predicted efficiency of transmission XRD versus x-ray energy for a 3 mm thick N5 nickel alloy specimen encased in a refractory oxide mold (6.4 mm wall thickness). The furnace ports are 10 mm thick borosilicate glass. The hot zone comprises an alumina core with 4.5 mm thick walls and molybdenum windings (1.6 mm diameter wire).

48

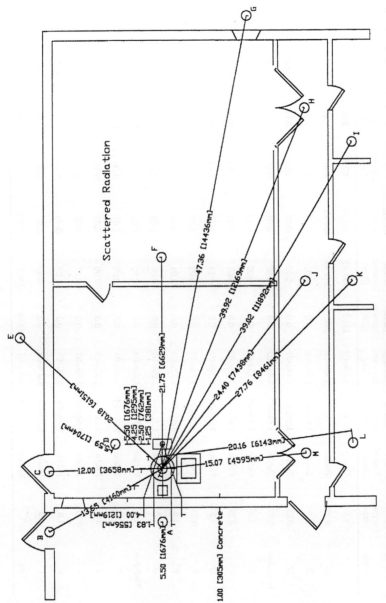

Figure 45. Drawing of the DS furnace room and adjacent areas, showing areas where radiation exposure levels were calculated and shielding design was considered.

Table 3. Calculated radiation exposure near the directional-solidification furnace during x-ray diffraction experiments.

Location	Occupancy Factor	Type of Area	Scatter Exposure (R/h)	Leakage Exposure (R/h)	Direct Exposure (R/h)	Total Exposure (R/hr)	Weekly Exposure (R)	Required TVL	Existing TVL	TVL Needed	Lead Required (mm)
A	1.0	Occup	0.	5.05e-02	231.37	231.42	1851.33	4.27	2.88	1.38	7
A'	0.25	Public	0.	5.05e-02	231.37	231.42	462.83	4.67	2.88	1.78	9
B	1.0	Occup	1.84e-03	1.44e-02	0.0	0.02	0.13	0.11	6.14	0.00	0
B'	0.25	Public	1.84e-03	1.44e-02	0.0	0.02	0.03	0.51	6.14	0.00	0
B window	0.25	Public	1.84e-03	1.44e-02	0.0	0.02	0.03	0.51	0.41	0.10	0
C	1.0	Occup	1.81e-03	2.13e-02	0.0	0.02	0.19	0.27	0.00	0.27	1
C'	0.25	Public	1.81e-03	2.13e-02	0.0	0.02	0.05	0.67	0.00	0.67	3
D	1.0	Occup	8.63e-03	1.29e-01	0.0	0.14	1.10	1.04	0.10	0.95	5
E	0.25	Public	7.23e-04	9.33e-03	0.0	0.01	0.02	0.30	4.31	0.00	0
F	1.0	Public	8.07e-04	8.72e-03	0.0	0.01	0.08	0.88	0.10	0.79	4
G	1.0	Public	1.71e-04	1.60e-03	0.0	0.00	0.01	0.15	3.04	0.00	0
G window	1.0	Public	1.71e-04	1.60e-03	0.0	0.00	0.01	0.15	0.29	0.00	0
H	0.25	Public	2.32e-04	2.28e-03	0.0	0.00	0.01	0.00	0.10	0.00	0
I	1.0	Public	2.28e-04	2.37e-03	0.0	0.00	0.02	0.32	0.34	0.00	0
J	0.25	Public	5.48e-04	6.39e-03	0.0	0.01	0.01	0.14	0.16	0.00	0
K	1.0	Public	3.90e-04	4.75e-03	0.0	0.01	0.04	0.61	0.27	0.35	2
L	1.0	Public	6.17e-04	8.07e-03	0.0	0.01	0.07	0.84	0.19	0.65	3
Test	1	Nobody	2.87e-02	3.00e-01	343.8	344.13	2753.03	5.44	2.18	3.26	16

5. DIFFRACTION EXPERIMENTS

Diffraction of high-energy x-rays was investigated experimentally, first by performing tests on well-characterized single-crystal specimens. Effects on the diffraction pattern of materials surrounding the specimen were studied and later, melting and solidification studies were performed. The first melting experiments on pure metals with low melting temperature demonstrated the feasibility of the diffraction technique. Next, aluminum and copper were studied in a medium-temperature test cell. Finally, the transmission x-ray diffraction method was tested on castings of nickel alloy in a directional solidification furnace.

5.1 Back-Reflection Laue Images

The first experiments verified that x-ray diffraction could be obtained using our equipment. We set up an apparatus typically used to form a Laue back reflection image. Our 160 kV tubehead was fitted with 1.5 mm diameter collimators, and the beam directed through an instant-process film cassette toward a piece of single-crystal turbine-blade casting. The source-to-specimen distance was 130 mm and the cassette was placed 30 mm from the specimen. With x-ray technique factors of 30 kV and 1 mA, the image shown in Figure 46 was formed after an exposure of 30 min.

Figure 46. Back-reflection Laue image (30 kV, 2 mA) of a single-crystal nickel-alloy casting specimen.

5.2 Reflection XRD Experiments

An x-ray beam, collimated to 2 mm diameter, was inclined at 45° with respect to one face of a single-crystal nickel-alloy casting (Fig. 47). The x-ray imager was also inclined at 45°, to give an angle of 90° (2θ) between the x-ray beam and imager. A definite x-ray diffraction pattern was obtained using a 50 kV x-ray tube voltage. The x-ray tube voltage was increased first to 75 kV, then to 100 kV and finally to 160 kV. The diffraction pattern persisted, with a very noticeable increase in intensity. The Compton-scattering background increased somewhat as tube voltage was increased. This experiment verified that x-ray diffraction could be obtained with tube voltages (160 kV) substantially higher than those used for conventional Laue measurements (40 kV).

We placed mold material 7 mm thick (typical of that used for turbine-blade castings) in the x-ray path during another reflection diffraction experiment. Figure 48 shows the recorded diffraction patterns without (a) and with (b) mold material. The diffraction spot intensity decreased with the mold present; however, the diffraction pattern was still discernable. No new diffraction spots arising from the mold were noted. Compton scatter, observable in the images of both the specmen alone and the specimen with mold material, diminished image contrast.

5.3 Transmission XRD

During reflection XRD experiments, the angles of incidence and scattering were varied to study the effects of angle on diffracted intensity. In one experiment on a thin single crystal of copper, a transmission geometry was used. The diffraction spots were of high intensity, and the Compton background was low. We quickly confirmed that a transmission geometry would permit probing the interior of an object.

Figure 47. Reflection x-ray diffraction experiment. The x-ray source, specimen, and imager are shown.

52

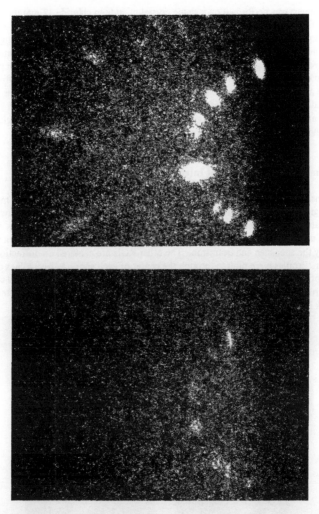

Figure 48. Reflection x-ray diffraction pattern (160 kV, 1 mA, $2\theta = 90°$) from (a) a nickel-alloy single-crystal specimen alone and (b) from the specimen, with a 7 mm thick piece of mold material placed between the x-ray source and the specimen.

5.3.1 Single-Crystal Specimen

An experiment was performed on a single crystal of sodium chloride 12 mm thick, to verify that transmission x-ray diffraction could be obtained on a thick specimen with high x-ray energies. The x-ray beam (320 kV, 3 mA), collimated to 1 mm in diameter, was directed along the [001] crystallographic axis. A source-specimen distance of 270 mm and a specimen-to-imager distance of 152 mm were used. Figure 49 shows the recorded transmission x-ray diffraction pattern. The central round dark area in the center is the shadow of the beamstop. The bright diffraction spots in the square array confirm the cubic structure of the sodium chloride crystal. Compare the image with the standard (001) projection of a cubic crystal shown in Figure 50. Although 256 video frames were averaged to achieve the high signal-to-noise ratio for this image, a single video frame still clearly showed the diffraction pattern.

5.3.2 Determining Diffraction Spot Intensity

The intensities of diffraction spots for the transmission XRD experiments were measured by defining a region of interest (ROI) surrounding each spot. Figure 51 shows the transmission diffraction pattern of a nickel-alloy single-crystal with four ROIs. ROI-2 is the area of transmission through the tungsten beamstop; it contains the primary broad-spectrum x-ray beam attenuated by the specimen and beamstop. The other ROIs are centered on diffraction spots from the specimen. Profiles were plotted through the center of each spot. Adjacent lines of pixels about the profile line were averaged to improve statistics. The maximum gray level was determined from the peak in the profile, and the background gray level was subtracted (Fig. 52) to give the peak intensity. A mean gray level was often computed as well.

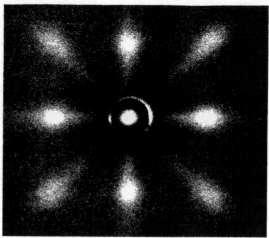

Figure 49. Transmission x-ray diffraction pattern (320 kV, 3 mA) from a 12 mm thick sodium chloride crystal. The x-ray beam was aligned along the [001] axis of the crystal.

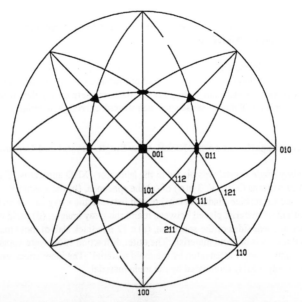

Figure 50. Standard (001) stereographic projection of a cubic crystal. The numbers are the Miller indices of the lattice planes.

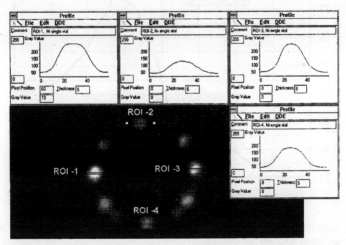

Figure 51. Transmission XRD image of the nickel-alloy single-crystal with four regions of interest (ROI) identified. Gray-level profiles through these ROIs are also shown.

5.3.3 XRD Experiments on a Single Crystal Surrounded by Amorphous, Textured, and Crystalline Materials

Transmission x-ray diffraction experiments were performed on a piece of single-crystal turbine-blade casting 6.2 mm thick. Since the goal our research was to investigate metals during the casting process, various materials were placed in the x-ray path during the diffraction experiment, to confirm the ability to sense specimen structure through materials external to it.

5.3.3.1 Diffraction Experiment on the Nickel-Alloy Single-Crystal and Ceramic Mold

The nickel-alloy single-crystal was placed in the laboratory XRD apparatus. Figure 53 shows transmission XRD patterns (320 kV, 3 mA) of (a) the specimen, (b) the specimen with 12 mm of mold material (used for turbine-blade castings) placed between the x-ray source and specimen, (c) the specimen with mold material placed between it and the x-ray imager, (d) mold material on both the entrance and exit sides of the specimen, (e) a 12 mm thick slab of mold material, and (f) the two 12 mm thick slabs of mold material. The outer diffraction spots were apparently of low energy and were thus strongly attenuated by the mold material. The inner spots, apparently of high energy, were only weakly attenuated by the mold material.

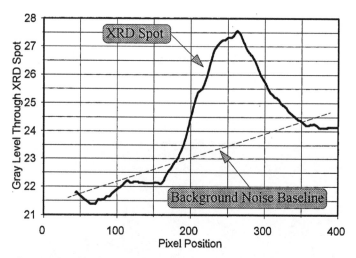

Figure 52. Gray-level profile through a diffraction spot showing the background noise level is generally not zero nor uniform.

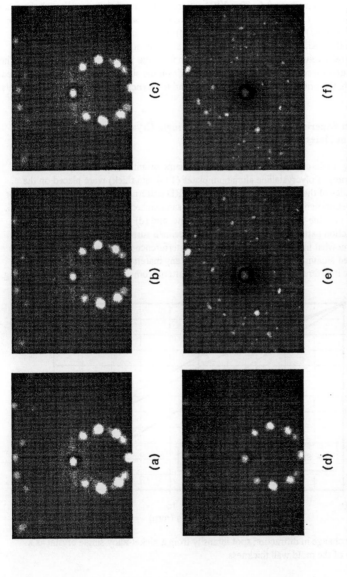

Figure 53. Transmission XRD patterns (320 kV, 3 mA) for combinations of a nickel-alloy single-crystal and mold material (typical of that used for turbine-blade castings). (a) the specimen, (b) the specimen with 12 mm of mold material (placed between the x-ray source and specimen, (c) the specimen with mold material placed between it and the x-ray imager, (d) mold material on both the entrance and exit sides of the specimen, (e) a 12 mm thick slab of mold material, and (f) the two 12 mm thick slabs of mold material.

57

Regions of interest were defined around each diffraction spot, and the spot intensity was measured as described in section 5.3.2. The relative change in spot intensity is plotted as a function of mold thickness in Figure 54. The two slabs of mold material only decreased the relative diffraction spot intensity to 0.6. This XRD experiment showed that even though the mold material may have a distinct diffraction pattern (arising from the coarse granular structure on the outside surface of the mold) the strong diffraction spots from the single-crystal nickel-alloy specimen dominate the XRD image. Even when the material surrounding the specimen is very coarse grained and relatively thick, the diffraction pattern from the specimen was observable. The brightness of mold and specimen diffraction spots depends on the sizes of crystallites encountered along the x-ray path, the orientation of the crystallites, and the structure factors of these materials.

5.3.3.2 Diffraction Experiment on the Nickel-Alloy Single-Crystal Surrounded by Fine-Grained Aluminum Plates

The nickel-alloy single-crystal was left undisturbed (same orientation with respect to the collimated x-ray beam). Polycrystalline aluminum plates (15.4 mm thick) were placed on the entrance and exit sides of the specimen. Transmission XRD patterns (Fig. 55) were recorded for (a) the nickel-alloy specimen alone, (b) for the specimen sandwiched between two 15.4 mm thick aluminum plates, (c) for the entrance aluminum plate alone, and (d) the both aluminum plates. Note that the diffraction pattern from the nickel-alloy specimen surrounded by the aluminum is only diminished somewhat in intensity, with little or no interference from the diffraction patterns from the fine-grained aluminum plates. Furnace walls of any material of low atomic number should provide little interference in detecting the XRD pattern from a single-crystal nickel casting.

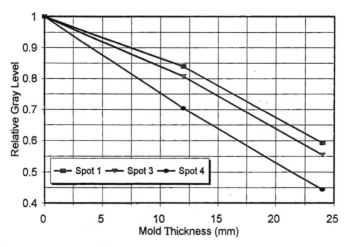

Figure 54. Relative change in diffraction spot intensity from a nickel-alloy single crystal as a function of the mold wall thickness.

58

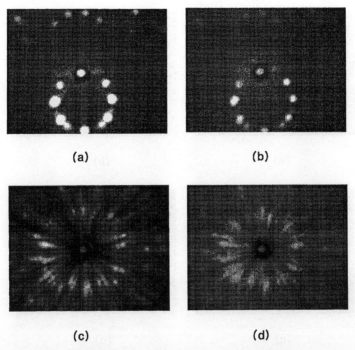

(a)   (b)

(c)   (d)

Figure 55. Transmission XRD patterns (320 kV, 3 mA) for combinations of a nickel-alloy single crystal and polycrystalline aluminum plates. (a) the nickel-alloy specimen alone, (b) for the specimen sandwiched between two 15.4 mm thick aluminum plates, (c) for the entrance aluminum plate alone, and (d) the both aluminum plates.

5.3.3.3 Diffraction Experiment on the Nickel-Alloy Single-Crystal Surrounded by Coarse-Grained Metal Material

Transmission x-ray diffraction was again performed on the nickel-alloy single-crystal. A coarse-grained aluminum rod 22 mm in diameter was placed on the entrance and the exit sides of the specimen, to determine the influence of surroundings having a distinct diffraction pattern on the XRD pattern of the nickel-alloy specimen. Figure 56 shows the four transmission diffraction patterns obtained from (a) the nickel-alloy specimen, (b) the specimen with aluminum rod placed on the entrance side, (c) the specimen with the aluminum rod placed on the exit side, and (d) aluminum rod alone. The transmission diffraction pattern of the nickel-alloy specimen still dominates the image; however, several high-intensity diffraction spots from the coarse-grained rod are evident. Coarse-grained material surrounding the specimen may cause some interference; however, the XRD pattern from the single-crystal nickel-alloy specimen is still clearly visible.

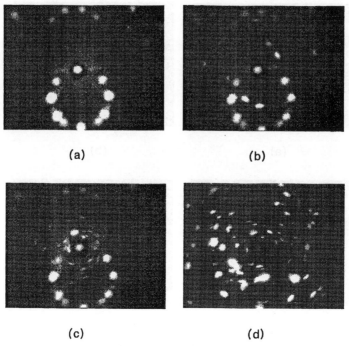

(a)   (b)

(c)   (d)

Figure 56. Transmission XRD patterns (320kV, 3 mA) for combinations of the nickel-alloy
single-crystal and a coarse-grained aluminum rod. (a) the nickel-alloy specimen, (b)
the specimen with aluminum rod placed on the entrance side, (c) the specimen with the
aluminum rod placed on the exit side, and (d) aluminum rod alone.

5.4 XRD to Determine the Physical State of a Specimen

5.4.1 Liquid-Solid Boundary in Gallium

Our investigations to demonstrate the feasibility of performing transmission XRD on a
specimen (contained in a mold) to determine its physical state (solid or liquid) began with a
gallium specimen. Its properties permitted an independent confirmation of the physical state of
the specimen. Gallium exhibits a change (3 percent) in density between the solid and liquid states,
large enough to produce a discernible difference in brightness on a radiographic image. The
ability to independently determine the position of the liquid-solid boundary (brighter versus darker
areas of the radiograph) provides a means for validating the spatial performance of the XRD
sensor.

The test cell shown in Figure 57 was used for the gallium experiments. A steady-state boundary was produced between the solid gallium (at the top because of its lower density) and the liquid gallium. The position of the probing x-ray beam could be moved into the solid or liquid. We could switch between XRD and radiographic imaging modes by removing the collimating apertures from the x-ray beam. Figure 58 shows radiographic and XRD images with the x-ray beam positioned (a) in the solid, (b) on the solid-liquid boundary, and (c) fully in the liquid gallium. The circular black area in the center of each radiographic image is the shadow cast by the tungsten beamstop. The x-ray beam is centered on the beamstop, so the dark area shows the position of the incident x-ray beam.

When the x-ray beam was positioned to probe only the solid gallium, bright diffraction spots were observed. In the liquid, a diffuse ring of scattering was visible. At intermediate locations between liquid and solid, the diffuse ring and diffraction spots were both present, but each was of diminished intensity.

5.4.2 Spatial Resolution Experiment

The temperature of the heater on the gallium test cell was adjusted to maintain a stationary position of the liquid-solid boundary. Transmission XRD images were recorded for a number of positions of the boundary with respect to the 1 mm diameter x-ray beam, by moving the gallium container vertically. The diffraction spot intensity was examined as a function of position of the x-ray beam relative to the liquid-solid interface. A region of interest was defined to enclose each diffraction spot. Spot intensity was determined as described previously.

Figure 57. Photograph of the gallium test cell.

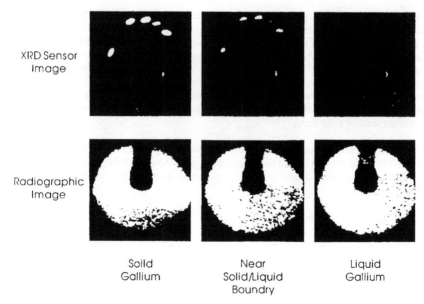

XRD Sensor
Image

Radiographic
Image

Solid
Gallium

Near
Solid/Liquid
Boundry

Liquid
Gallium

Figure 58. Radiographic images and transmission XRD images with the x-ray beam probing the
solid (top), the solid/liquid boundary (middle), and the liquid (bottom). The upper
portion of the radiographic image is brighter, because the x-rays are less attenuated by
the lower-density solid gallium.

Figure 59 shows a plot of the relative intensity of the brightest diffraction spot as a function of
vertical position. The position of the x-ray beam ranges from fully.in the solid (0 mm) to a
position fully in the liquid (0.8 mm). The plot indicates that the XRD technique has a spatial
resolution (0.8 mm) roughly equal to the diameter of the probing x-ray beam (1 mm). The profile
plotted in Figure 59 is the convolution of the the sensed area (a small circular x-ray beam) with
the actual physical boundary. The resolution of the sensed position may be improved by
deconvolution to remove the effects caused by the finite-size of the x-ray beam.

5.4.3 Diffraction Experiments with the Energy-Sensitive Detector

Transmission diffraction experiments were also performed with the germanium (energy-
sensitive) detector in place of the x-ray imager. The much higher efficiency of the germanium
detector, compared with the x-ray imager, required replacement of the collimating apertures. The
1 mm diameter aperture, used in experiments with the x-ray imager, produced a diffracted beam
too intense for the detector electronics. An aperture 0.2 mm in diameter was used.

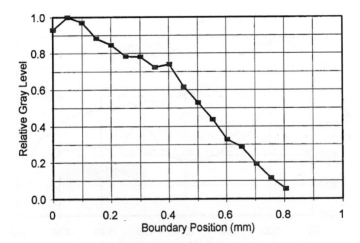

Figure 59. Relative intensity of one diffraction spot as a function of the displacement as the x-ray beam was moved with respect to the liquid-solid boundary in the gallium test cell.

Figure 60 shows transmission spectra recorded for acquisition times of 30 s. A source-specimen distance of 250 mm and a specimen-detector spacing of 180 mm was used for these experiments. The detector collimator was not used for these experiments. The spectra were obtained with an x-ray tube potential of 160 kV and a tube current of 1 mA. The energy spectra below 90 keV are not shown because of the complexity introduced by the characteristic x-ray peaks from the tungsten x-ray source and fluorescence radiation from the lead collimating apertures on the x-ray source.

The spectral peaks at approximately 100 keV and 130 keV are produced by x-ray diffraction spots from the solid gallium. There were no discernible peaks in the spectrum recorded with the liquid gallium. Although the intensity was higher for the diffraction spots, the spots were highly localized. The spectrum for the liquid specimen records more counts in each energy interval, although the intensity is lower, because spatially it is a very large ring, which illuminates more of the detector.

While the gallium fixture was in place, we took the opportunity to investigate the effects of mold material when recording the diffracted energy spectrum. Figure 61 shows the measured transmission diffraction spectrum when a piece of ceramic mold material 10 mm thick was placed on the exit side of the gallium specimen. The spectra show that the transmitted intensity is decreased by the mold, but that the spectral shape is changed very little.

63

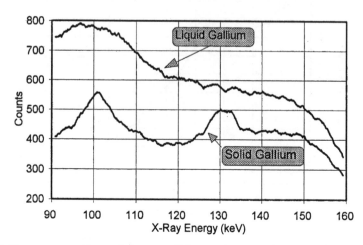

Figure 60. Energy spectra of transmission x-ray diffraction from a single crystal of gallium and from molten gallium.

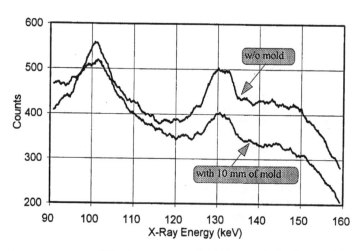

Figure 61. Transmission x-ray diffraction spectrum of a single-crystal of gallium and the spectrum when a piece of mold material 10 mm thick was placed in front of the gallium.

5.4.3.1 Preliminary Experiments to Study Fraction of Solid Along the X-Ray Beam Path

The effects of varying the path lengths of liquid and solid metal encountered by the probing x-ray beam were studied by constructing a dual step-wedge specimen (Fig. 62). Acrylic plastic was machined to produce two step-wedge cavities. The cavities were filled with gallium metal. One of the step wedges was electrically heated to melt the gallium. The other step-wedge, thermally isolated by the acrylic plastic, remained in a solid form. With the two wedges aligned, the x-ray path length through gallium remains the same (14 mm) for each step; however, the percentages of liquid and solid vary. The x-ray path length through the acrylic plastic is identical for each set of steps.

The step wedge that was to contain solid gallium was melted and then cooled slowly to produce a single-crystal. Transmission XRD images of the solidified gallium in this step wedge showed the same spatial position of diffraction spots for each of the steps, indicating a single-crystal specimen had been produced.

An x-ray beam (160 kV, 1 mA) 1 mm in diameter was directed through the two wedges (one liquid and the other solid). Figure 63 shows the arrangement of x-ray source, collimator, step-wedge, beamstop, and germanium detector. Detector shielding (not shown in the figure) was necessary to prevent stray radiation (leakage from the tube housing and scatter from the room) from corrupting the measured transmission spectra from the gallium wedges. Transmission spectra (Fig. 64) were recorded (30 s live time) for each set of steps by moving the step-wedge vertically. The broad diffraction peak at 130 keV begins to disappear as the x-ray beam probes regions with decreasing amounts of crystalline solid. Measurement of the fraction of solid from the diffraction spectrum appeared feasible.

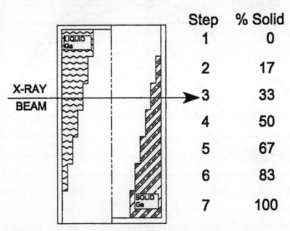

Figure 62. Drawing of the dual step wedge test specimen used for studies of diffraction from differing relative path lengths through liquid and solid.

Figure 63. Photograph of the apparatus used for energy-sensitive transmission diffraction experiments on a dual step-wedge gallium specimen.

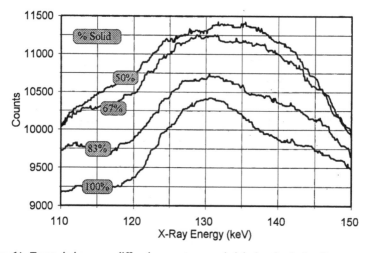

Figure 64. Transmission x-ray diffraction spectra recorded during the dual gallium step-wedge experiments.

5.5 Medium Temperature (<1000 °C) Furnace Experiments

Figure 65 illustrates the key elements in the apparatus used for transmission XRD measurements during melting and solidification of aluminum and copper. The x-ray tube, with supplemental lead shielding on the tube housing, is on the right side of the picture. The collimated beam (1 mm in diameter) passes through the furnace (white cylinder in the center), through a quartz specimen tube, and into the metal specimen. A portion of the x-rays entering the specimen is diffracted. The x-ray imager (black cylindrical object on the left) detects and displays the x-ray diffraction pattern. The tungsten beam-stop is mounted between the furnace and the x-ray imager.

5.5.1 Aluminum Melting Experiments

A melting and recrystallization experiment was performed on an aluminum specimen in the gradient furnace. A 22 mm diameter, coarse grained polycrystalline, 99.999 percent pure aluminum rod, in a quartz tube (2.2 mm wall thickness, 27 mm i.d.), was placed in the furnace. Figure 66 shows the transmission Laue x-ray diffraction patterns obtained during heating, melting, and then resolidification of the aluminum. As the rod was heated, the diffraction pattern changed, reflecting changes in the physical state of the specimen. The complex diffraction patterns in the first few frames of the image sequence are the result of interactions of x-rays with many crystals in the polycrystalline specimen. As the temperature was raised further, the larger grains in the polycrystalline rod grew at the expense of the smaller ones (Ostwald ripening), and the diffraction

Figure 65. Apparatus for observing transmission diffraction during melting and solidification of metal.

67

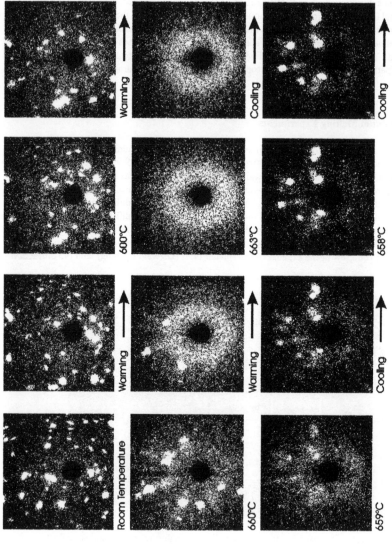

Figure 66. Sequence of transmission Laue diffraction patterns (160 kV, 1 mA) obtained as a polycrystalline aluminum rod was heated, melted, and then cooled.

pattern became simpler as the x-ray beam encountered fewer, but larger, crystals. Near the melting point (652°C), the diffraction pattern began to lose order. The Laue pattern degenerated into a diffuse ring of scattering when the aluminum was fully melted. The difference between the patterns produced by the solid aluminum and the liquid aluminum was dramatic and unmistakable. The Laue diffraction spots disappeared at the same time the diffuse ring formed. As the aluminum cooled below the melting point, the diffraction spots reappeared at the same time that the diffuse ring disappeared. The diffraction pattern for the resolidified aluminum specimen was coarser, indicating an x-ray path passing through a few, large grains, or grains favorably oriented to diffract the intense characteristic radiation from the tungsten-target x-ray source.

In an identical heating experiment, the polycrystalline aluminum rod was heated to a temperature of only 610°C, where grain growth would occur without melting. The specimen was then cooled and sectioned for analysis. Figure 67 shows polished and etched cross sections of the virgin polycrystalline rod and the annealed rod, confirming that the initial reduction in the number of diffraction spots during the initial portion of the experiment was a result of grain growth.

Figure 67. Micrographs of polished and etched specimens from an aluminum melting experiment. Left: polycrystalline rod, before melting. Right: specimen after annealing.

5.5.2 Copper Melting Experiment

A polycrystalline copper rod 12.7 mm in diameter was placed in a quartz tube (2.2 mm wall thickness), which had a triangular cross section. The tube was then inserted into our gradient furnace. A collimated x-ray beam 1 mm in diameter was directed into the furnace, and the real-time x-ray imager was placed on the opposite side of the furnace to record the transmission diffraction pattern. The x-ray technique factors were 160 kV and 1 mA.

The entire furnace could be manipulated remotely to move it vertically and horizontally. The vertical movement was used to scan the x-ray beam and imager with respect to the liquid-solid copper boundary that can be established in the gradient furnace. Horizontal movement across the wedge-shaped copper specimen permitted interrogating different thicknesses.

69

Figure 68 shows a collage of XRD images recorded as the temperature of the furnace was raised, to melt the copper, and then lowered, to solidify it. The sequence of warming and cooling was repeated several times. As in the aluminum melting experiments, the solid copper produced diffraction images with many bright Laue diffraction spots. When the melting temperature of copper was exceeded, the ordered diffraction pattern disappeared and was replaced by a diffuse ring of x-ray scattering from the molten copper. The diffuse scattering ring was observed in many of the diffraction images. This was presumably the result of partial melting of the copper rod near the outside surface of the specimen. Rapid heating was not possible because the upper temperature limit of the furnace was close to the melting temperature of copper.

5.6 Sensing Solidification During Investment Casting

A directional solidification furnace, used for prototype single-crystal turbine blade castings, was fitted with an x-ray source and real-time x-ray imager as shown in Figure 69. X-rays are collimated and directed through a borosilicate glass entrance port into the vacuum furnace. The x-ray beam passed through bats of fiberous cermaic insulation, through the molybdenum furnace windings, through the alumina core of the hot zone of the furnace, through the casting mold, and into the nickel-alloy specimen. Diffracted x-rays from the specimen were transmitted through the remaining portion of the specimen and the exit wall of the mold. Attenuation along the exit path of the diffracted x-rays through alumina, molybdenum, insulation, and borosilicate glass port reduced the intensity of the diffraction spots. No beamstop was required because the intensity of the primary beam was not high enough to damage the imager.

Before the casting experiments, tests were performed to determine whether diffraction from furnace components would interfere with the patterns produced by a casting. The 10 mm thick borosilicate glass ports were placed in our laboratory XRD apparatus (253 mm source-specimen distance, and 381 mm source imager distance). The diffraction patterns are shown in Figure 70. Profiles through these images are shown in Figure 71. The relatively featureless diffraction patterns from the amorphous glass should not interfere with diffraction from a metal specimen in the furnace. The pattern does not introduce new spots, but does reduce intensity.

The borosilicate glass ports of the directional solidification furnace attenuate both the incident and diffracted x-rays. A single-crystal N5 casting, in its mold, was placed in the unheated furnace. The x-ray source (320 kV and 3 mA) and x-ray imager were positioned, and specimen rotation adjusted, to obtain three strong diffraction spots from the alloy casting. XRD images were recorded for three cases: (a) no ports on the furnace, (b) one, and then (c) both glass ports attached to the furnace. Figure 72 shows the relative transmission of the primary beam and of one of the diffraction spots (approximately 200 keV energy) for the three configurations of the ports. This experiment, confirmed by modeling, revealed that a significant improvement in diffraction spot intensity could be achieved by replacing the glass ports with a lower-attenuation material. As discussed in section 4, replacement of the glass ports with the graphite/epoxy ports increased signal-to-noise ratio by a factor of 1.5.

70

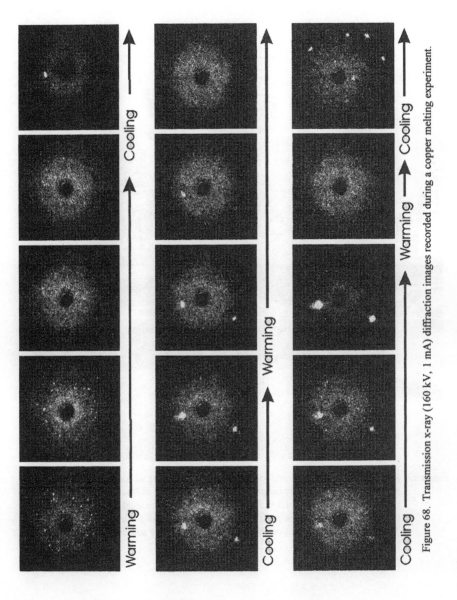

Figure 68. Transmission x-ray (160 kV, 1 mA) diffraction images recorded during a copper melting experiment.

Figure 69. Resistively heated directional solidification furnace, fitted with a collimated x-ray source and real-time x-ray imager for transmission x-ray diffraction studies during casting of single-crystal nickel superalloys. The radiation shielding enclosures have been removed to show the x-ray source (on the right) and the imager (on the left). The motion control stages can be used to move the x-ray source and imager laterally or vertically.

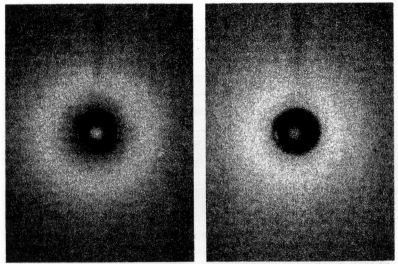

Figure 70. X-ray diffraction patterns (320 kV, 3 mA) recorded for one (left) and two (right) of the borosilicate glass ports from the directional solidification furnace.

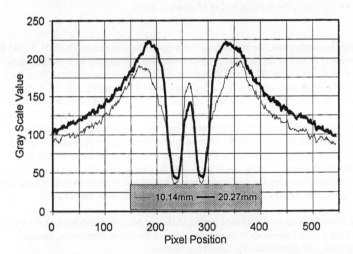

Figure 71. Gray-level profiles through the XRD patterns from the borosilicate glass furnace ports. The single broad peak in the radial distributions is characteristic of diffraction from an amorphous material.

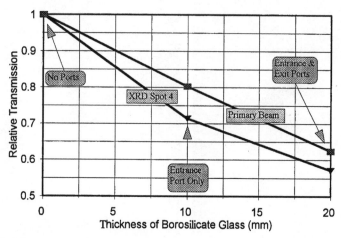

Figure 72. Measurements of x-ray transmission through the 10 mm thick borosilicate glass ports of the directional solidification furnace. The plots are normalized to the value of transmission with no ports attached to the furnace. A mold-encased N5 casting (3 mm thick) was in the furnace during all measurements.

In another laboratory test, we recordeded the x-ray diffraction pattern (320kV, 3 mA) from a hot zone identical to that in the furnace. The hot zone was constructed with molybdenum windings on an alumina core. Figure 73 shows the diffraction patterns recorded from the hot zone (top) and the hot zone with a nickel-alloy single-crystal 6.2 mm thick inside it (bottom). There were some x-ray diffraction spots of low-intensity from the molybdenum windings in the contrast-enhanced image of the hot zone alone (left). The most striking feature of the image on the left is the shadow of the molybdenum heater windings. Even though the x-ray beam was collimated, Compton scattering from the alumina core of the hot zone produced x-rays of enough intensity to form a radiographic image of the hot zone. Placing the nickel specimen in the hot zone severely attenuates the scatter which produced the radiographic artifacts in the left image. The nickel specimen becomes the dominant x-ray diffractor and produces a ring of high-intensity diffraction spots, with minimal interference from the interactions with the hot zone.

These laboratory tests on the furnace ports and the hot zone demonstrate that the x-ray diffraction from a nickel specimen of moderate thickness should be observable, despite interference from, and attenuation by, furnace components.

74

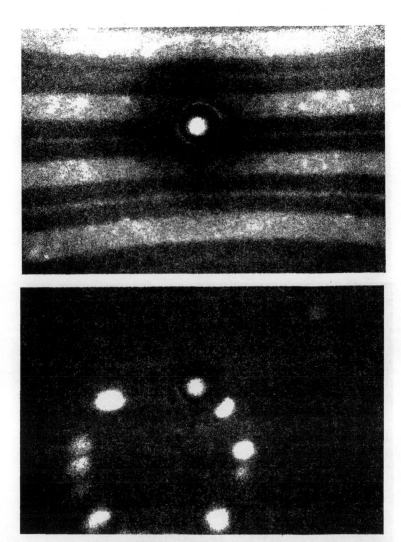

Figure 73. X-ray diffraction patterns from both the hot zone (molybdenum windings on an alumina core) (top) and the hot zone with a nickel-alloy single-crystal in it (bottom). Scatter from the alumina furnace core broadens the x-ray beam and forms a radiographic-type image overlay to the diffraction pattern. The dark bands in the top image are the result of higher attenuation in the molybdenum windings.

75

5.6.1 XRD Tests in the Directional Solidification Furnace

Before heating the furnace, we placed the 6.2 mm nickel-alloy single-crystal (on a rotational stage) into the center of the furnace. The x-ray system was adjusted for a tube voltage of 320 kV and a tube current of 3 mA. X-ray diffraction images were recorded as the specimen was rotated in the collimated x-ray beam. Figure 74 shows XRD images of the specimen within the furnace, for two different rotations of the specimen. The bright spot at the right edge of the images is the primary x-ray beam. No beamstop was used.

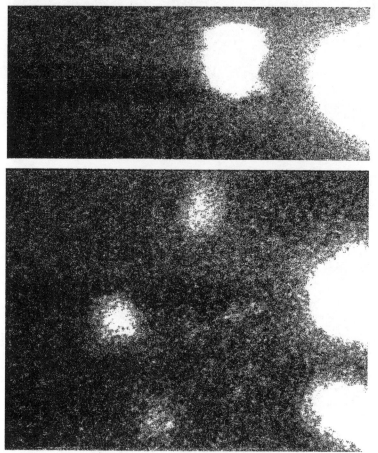

Figure 74. XRD images from a nickel-alloy single crystal 6.2 mm thick placed in the directional solidification furnace. The two images were recorded for different rotations of the specimen.

An experiment was performed to identify the lattice planes responsible for particular diffraction spots. A previously-cast, nickel-alloy single-crystal 3 mm thick in a mold (5 mm wall thickness) was placed in the directional solidification furnace. The collimated x-ray beam (320 kV, 3 mA) 1 mm in diameter produced x-ray diffraction patterns similar to that shown in Fig. 74. The x-ray imager was translated to place the primary x-ray beam at the center of the field of view of the imager. The position of the translation stages was noted. Various diffraction spots were brought into view by rotating the specimen about the vertical axis. For each spot, the translation stages of the imager were used to position the particular diffraction spot at the center of the imager. The position of the stages was recorded. From the two sets of translation stage positions (primary beam and diffraction spots), we were able to determine the position of the diffraction spot relative to the primary beam. A diffracted angle, 2θ, was computed for each diffraction spot (arctangent of the spot position with respect to the primary beam divided by the 425 mm specimen to imager distance). The energy of each diffraction spot was estimated by decreasing the x-ray tube potential until the spot vanished. Tube current was adjusted to compensate for the intensity changes produced as the tube potential was varied. A lattice spacing was computed (Bragg Law) from the scattering angle and the x-ray wavelength. Table 4 contains the data and computed parameters. The lattice spacings fall into two groups. The group clustered around 1.277 Å represent reflections from the (220), (440), (660), etc. planes (lattice spacing for the 220 planes in nickel is 1.24592). The second group, clustered around 2.0853 Å, are reflections from the (111), (222), (333), etc. planes (lattice spacing for the 111 planes in nickel is 2.03458).

Table 4.. Calculation of lattice parameters from XRD data.

Spot ID	Distance (mm) from primary beam to spot	2θ (°)	Tube potential (kV) when spot extinguishes	Wavelength (Å) when spot extinguishes	Computed lattice spacing $\lambda/[2\sin\theta]$ (Å)
1	28.47	3.83	144	0.0861	1.288
2	22.15	2.98	180	0.0689	1.323
3	21.37	2.88	180	0.0689	1.371
4	29.84	4.02	145	0.0855	1.220
5	34.84	4.69	128	0.0969	1.185
6	37.02	4.98	112	0.1107	1.275
7	14.82	2 .00	170	0.0729	2.093
8	12.15	1.64	194	0.0639	2.236
9	24.02	3.24	114	0.1088	1.927

5.6.2 Experiments with the Germanium Detector

The efficiency of our 13 mm thick germanium energy-sensitive detector is quite high, making it attractive for detecting diffracted beams of high energy. As discussed in section 2, x-ray diffraction is manifested spatially as spots, or as peaks in the transmission energy spectrum. A diffraction spot contains one energy or energies related by integer multiples (diffraction orders). Several experiments were performed with our germanium detector replacing the imager as an x-ray detector. A lead collimator (25 mm thick, 3.5 mm diameter hole) could be affixed to the detector, permitting small areas of the x-ray diffraction field to be probed.

A previously cast single-crystal of N5 alloy 6 mm thick in a mold of 5 mm wall thickness was placed in the unheated directional solidification furnace. This casting, which had a wedge shaped cross section, permitted investigation of different thicknesses of the casting by moving our x-ray source and detector laterally. Figure 75 shows the x-ray diffraction image recorded for this experiment. After the image was acquired, the x-ray imager was replaced by the germanium detector. Figure 76 plots transmission spectra with the germanium detector, positioned at the center of the primary beam or at the center of diffraction spot 4 (the one to the left of the primary beam). The small spectral peaks in the 70 to 90 keV range are K-characteristic x-rays lines from the lead collimators on the x-ray source and detector. The prominent spectral peak at 180 keV is produced by x-ray diffraction in the N5 specimen.

Without moving either the specimen or the x-ray source, the germanium detector was translated laterally (to vary the angle where diffraction was measured) in increments of 2 mm across the XRD spot. A spectrum was acquired at each position. Figure 77 shows that the peak energy is a smooth function of the diffraction angle from the specimen. As the diffracted angle is altered, the energy required to satisfy the Bragg equation changes. The difference in the height of the spectral peaks results from changes in the intensity of the incident x-ray spectrum at these energies and the efficiency of transmission x-ray diffraction. Refer to our model for transmission diffraction (section 3), which shows changes in efficiency of diffraction as a function of x-ray energy.

5.6.3 Producing Single-Crystal Castings

The directional solidification furnace was used to produce several single-crystal test castings before beginning the XRD experiments. A mold to produce a cylindrical casting was placed in the furnace, and a nickel-alloy charge placed in the crucible at the top of the mold. The charge was melted and a slow withdrawal initiated in an attempt to cast a single-crystal. After casting, the specimen was chemically etched to disclose the crystalline structure. No crystal boundaries were apparent. The high-energy transmission XRD system was used to evaluate the structure of the casting. The diffraction pattern from the cylinder (Fig. 78) clearly indicates that it is a single-crystal.

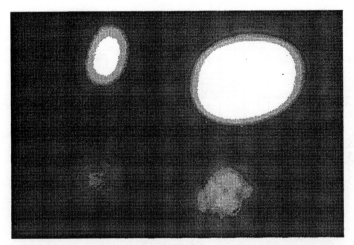

\Figure 75.   Transmission XRD image (320 kV, 0.75 mA) of a mold-encased N5 specimen placed in the directional solidification furnace. The bright area at the upper right is the primary x-ray beam. There are three XRD spots, to the left and below the primary beam.

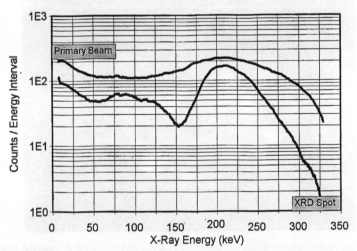

Figure 76. Transmission spectra with the collimated germanium detector positioned at the center of the primary beam or at the center of diffraction spot 4.

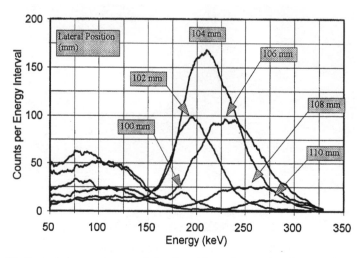

Figure 77. Transmission spectra recorded as the collimated germanium detector was translated laterally across the diffraction spot to the left of the primary, in increments of 2 mm.

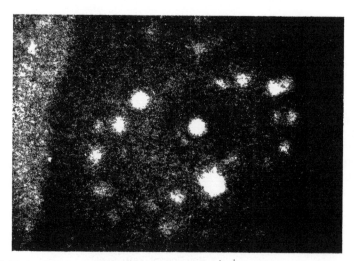

Figure 78. Transmission x-ray diffraction pattern (320 kV, 3 mA) obtained through the center of the cylindrical nickel-alloy specimen (15.6 mm in diameter) which was cast in the directional solidification furnace.

5.6.4 XRD During Investment Casting

A charge of N5 nickel-alloy was placed in the crucible at the top of the flat bar mold (3 mm thick by 38 mm wide cavity), the mold placed on the ram of the furnace, and the procedures listed in Table 5 followed to cast a single-crystal specimen.

Table 5. Procedures for operating the directional solidification furnace.

Step	Procedure
1	Open the gate valve between the load chamber and the furnace and use the roughing and diffusion pumps to produce a 1 Pa vacuum
2	Close the gate valve to the load chamber, bring the chamber to atmospheric pressure
3	Load the alloy charge into the crucible at the top of the mold and place the mold on the vertical ram
4	Close the gate valve, rough pump the load chamber to 3 Pa, and then open the gate valve
5	Ramp (over 30 min) both the upper and lower portions of the hot zone to a temperature of 260 °C (500 °F)
6	Ramp (over 30 min) both the upper and lower portions of the hot zone to a temperature of 538 °C (1000 °F)
7	Ramp (over 30 min) both the upper and lower portions of the hot zone to a temperature of 1093 °C (2000 °F)
8	Slowly raise (over 30 min) the mold into the hot zone, pausing if the vacuum deteriorates due to outgassing from the mold
9	Dwell for 45 min
10	Rapidly ramp (over 10 min) both the upper and lower portions of the hot zone to a temperature of 1566 °C (2850 °F)
11	Dwell for 30 min
12	Withdraw the casting from the hot zone, at a rate of 150 mm/h (6 in/h).

By removing the collimator of the x-ray source, we could form a radiographic image of the mold, with an approximately 30 mm field of view. Figure 79 (left) shows the grain selector of the bar mold before any N5 had melted. Figure 79 (right) shows a radiographic image of the grain selector immediately after it had been filled with molten N5 alloy. We radiographically observed filling of the bar mold with molten N5 material. After all the N5 had melted, the furnace ram was used to slowly withdraw the mold through the temperature gradient established between the hot zone and the cold plate in the furnace.

The collimator was placed on the x-ray tube, and the x-ray source and x-ray imager were scanned in unison over the mold in an attempt to observe x-ray diffraction spots from the solidifying N5 alloy. The x-ray techniques factors were 320 kV and 3 mA. No diffraction spots from the solid were found. Apparently, the solidification front was much lower than expected, and was below our field of view. The withdrawal of the mold from the furnace was completed, and the N5 was allowed to cool.

A second N5 casting experiment was performed. This time the mold with the wedge-shaped cross section was used. We thought that a greater thickness of N5 would lead to more intense (and hence more easily observable) diffraction spots.

The collimated x-ray source and imager were scanned to the thin edge of the wedge and then across it to thicker parts of the wedge. No diffraction spots were apparent. We suspected that the solidification of the N5 was again occurring below our field of view. The x-ray beam was positioned toward the thin edge of the wedge, the hot zone of the furnace was turned off, and we observed the x-ray diffraction pattern as the N5 began to cool and solidify. A diffraction spot began to appear in the center of the image, just above the primary x-ray beam.

At this point modifications were made to the furnace to raise (with an alumina foam spacer) the location where solidification occurred into the field of view of the XRD system. We also replaced the glass furnace ports with graphite-epoxy ports and changed the scintillator on the x-ray imager. The details of the changes made to the furnace, and to the x-ray imager were detailed in section 4 of this report. The brightness of the XRD spots increased by approximately an order of magnitude when the ports and scintillator were changed.

In the next experiments, the x-ray beam was directed into the base of the starter block of the casting just above the water-cooled ram of the directional solidification furnace. Solidification of elongated polycrystals was expected in this region. Several weak XRD spots were observed. Rotation of the casting brought the spots into and out of view of the x-ray imager. We used the motion stages on the x-ray source and imager to scan vertically into the funnel-shaped region of the mold. Initially, when this area was liquid, no diffraction spots were observed. As the solidification progressed upward, the funnel-shaped area began to solidify and diffraction spots began to form. Transmission XRD spots were observable when the x-ray beam was directed into a solidified region of the casting. Generally, only one diffraction spot was observed at a time, although others could be brought into the small field-of-view of the x-ray imager, by rotating the casting.

82

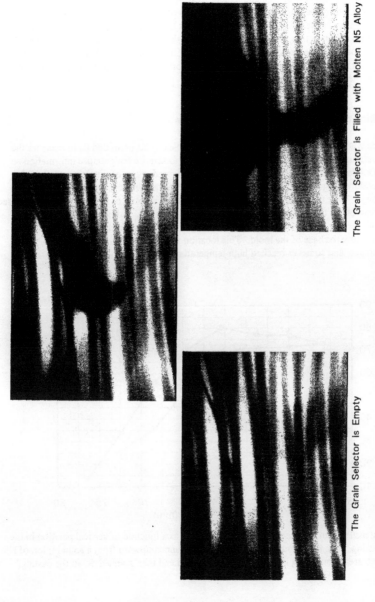

The Grain Selector is Empty

The Grain Selector is Filled with Molten N5 Alloy

Figure 79. Radiographic images of the grain selector of the bar mold (left) before, (middle) during, and (right) after filling with molten N5 alloy.

83

We optimized (by rotating the specimen) the intensity of a particular diffraction spot and used the motion stages on the x-ray source and imager to scan the probing x-ray beam vertically. The diffraction spot intensity was highest in the lower (solid) portion of the casting and gradually decreased in intensity as the XRD sensor probed regions higher (more liquid) in the casting. Repeated vertical scans, both up and down (Figure 80), showed the same behavior in intensity increase (more solid) and decrease (more liquid) in the diffraction spot. Other diffraction spots were rotated into view and additional vertical scans performed. The intensity of these diffraction spots varied in the same manner.

5.6.3.1 Probing the Dendritic Region of Solidification

Bar molds were fabricated with holes for thermocouples. This permitted us to measure the vertical temperature profile in the casting. The plan was to acquire temperature information so that our XRD data could be quantitatively compared with predictions of the fraction of solid versus temperature. Figure 81 is a drawing of the mold, with dimensions and thermocouple locations shown. Type C thermocouples were used. A small amount of alumina cement was used to affix the bead of the thermocouple in the hole through the mold. Additional cement was applied to seal the hole. The length of the thermocouple wires was trimmed so they reached about 50 mm below the base of the mold. This location was below the top of the water-cooled ram of the furnace, and so never reached high temperature.

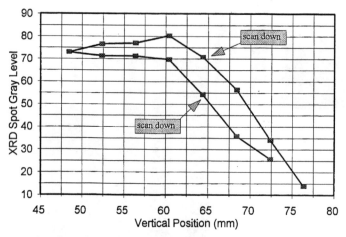

Figure 80. Intensity of an intense diffraction spot plotted as a function of vertical position in the casting. The x-ray source and imager were scanned upward from a solid region of N5 into areas containing less solid and more liquid and then scanned down the casting.

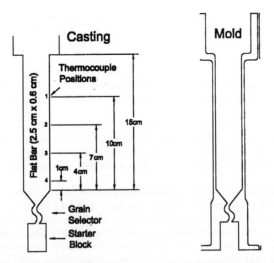

Figure 81. Drawing of the mold, which had provisions for placement of thermocouples. Holes were left in the mold so four thermocouples could be inserted and cemented in place.

In several of the casting experiments, a woven ceramic sheathing was used to cover the bare thermocouple wire. This sheathing was replaced with alumina tubes in later experiments. Extension wire was welded to the thermocouples and connected to an insulated feedthrough on the load chamber of the furnace. The extension wires were slightly coiled to prevent kinking as the mold (with attached thermocouples) was raised into the hot zone of the furnace. Software, written for a computer-based acquisition board, was used to periodically read the temperature of the thermocouples and write the data to a file.

The molds were fabricated from a refractory oxide mixture to produce bar-shaped castings 6 mm thick and 25 mm wide. A 350 gram charge of N5 was placed in the crucible at the top of the mold, and the mold placed on the furnace ram. The procedures given in Table 5 were followed. A few minutes after the end of the the final temperature ramp to 1566 °C, droplets of molten N5 were observed, filling the starter block in the mold. The remainder of the mold filled with N5 a few minutes later. After allowing the temperature of the mold and alloy to stabilize at for 30 min, the mold was slowly withdrawn from the hot zone at a rate of 150 mm/h. Solidification began at the base of the starter block, which rests on the water-cooled ram, and proceeded upward into the grain selector as the mold was withdrawn from the furnace. The structure of the casting from the grain selector upward was that of a single crystal. Polished and etched sections of the completed casting showed a dendritic structure with several phases and an absence of grain boundaries.

The thermocouple temperature data were well behaved (Figure 82) until the end of the last temperature ramp to 1566 °C. At this temperature, the data became extremely noisy. The electrical power to the hot zone of the furnace was turned off, to determine if noise pickup from the heating element power supplies caused problems in the thermocouple readings. The effect of turning off the furnace power was minimal. During several experiments, output from one or more thermocouples was lost. Noise on the remaining thermocouples remained high. The erroneous readings from thermocouples and the noise in all readings precluded using the temperature data as planned. We tried to remedy the temperature measurement difficulties, but did not succeed.

Filling of the mold was observed radiographically by removing the collimator (lead apertures) from our x-ray source. A field of view 31 mm in diameter could be observed in the casting with our 50 mm diameter x-ray imager. X-ray technique factors of 180 kV, 1 mA (1.2 mm diameter focal spot) produced acceptable images of the casting within the hot zone of the furnace.

The x-ray beam and imager were then lowered (using the remotely controlled motion stages) to the alumina foam spacer beneath the hot zone. The foam provides a low-loss path for x-rays and a clear field of view for the XRD solidification experiments. We waited until the casting had withdrawn to the point where the x-ray beam was centered in the funnel-shaped area of the casting, just above the grain selector. The collimator was placed on the x-ray source to configure the system to observe x-ray diffraction, and the tube potential and current raised to 320 kV and 3 mA. The higher tube voltage and current were necessary to produce x-rays able to penetrate through both the mold and the alloy casting.

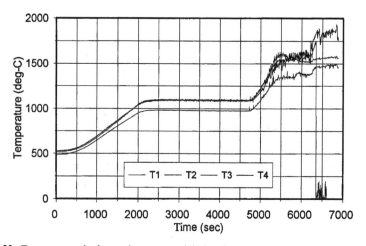

Figure 82· Temperatures in the casting measured during directional solidification.

The casting was manually rotated, by turning the ram, while observing the real-time transmission x-ray image. Before the solidification front advanced to the region probed by the x-ray beam, only the primary beam was visible. Eventually, solidification proceeded into the area probed by the x-ray beam and diffraction spots appeared. Rotating the specimen caused the spots to appear, move along hyperbolic paths through the center of the primary beam, and then disappear. The spots were produced as different lattice planes in the single-crystal casting satisfied the conditions for Bragg reflection. Since we were using a white beam source, there was not a singular angle where a spot appeared, but rather there was a small range of specimen orientations over which a spot was produced. The highest intensity spot was obtained when the peak intensity in the bremstrahlung spectrum from the x-ray source had the correct wavelength for the particular Bragg reflection.

The single-crystal structure in a casting could be confirmed by spatially scanning the x-ray source and imager. If the diffraction pattern remained the same, the region scanned possessed a single-crystal structure. After confirming crystal structure of a solidified portion of the casting, the x-ray source and imager were scanned vertically from the crystalline (solid) region through the region of dendritic solidification into a fully molten region of the casting. During the scan the XRD spot intensity decreased, as shown in Fig. 80. During similar tests, we acquired an extensive set of data.

The raw intensity data must be processed to account for the physics of the diffraction interaction and the spatial smoothing produced by an x-ray beam of finite-size scanned past a moving solidification front. As the x-rays pass through the region of dendritic solidification, their path is partially solid and partially liquid. Interaction of the x-rays with the crystalline solid produces diffraction spots. Interactions with the liquid does not contribute to spot intensity. Rather, it attenuates the incident x-ray beam and the beam diffracted from the solid, and contributes to a diffuse diffraction ring. As described in section 3, the spot intensity varies linearly with the fraction of solid.

A model (Lever) prediction of the fraction of solid versus temperature for N5 alloy is plotted in Figure 83 along with the measured gray level of the transmission diffraction spot. Since the temperature profile in the casting was not measurable in this experiment, we have adjusted the vertical and horizontal scale of the XRD data to coincide with the solidification model predictions. It is encouraging that the shape of the modeled and experimental curves are similar. The experimental data also appear to show formation of the predicted second phase (break in the curve).

5.6.3.2 X-Ray Topography of Dendrites During Solidification

In one of the casting experiments, the single crystal [001] axis was exactly aligned with the vertical axis of the furnace ram. With the x-ray beam slightly angled with respect to the (001) planes, a strong Bragg reflection was observed directly above the primary beam. The spot remained, but its mottled appearance changed as the specimen was rotated about a vertical axis.

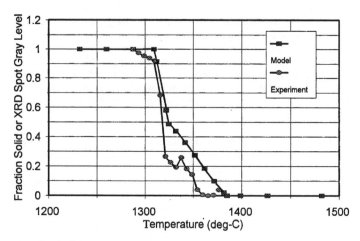

Figure 83. Gray level of the transmission XRD spot plotted along with model (Lever) predictions of the fraction of solid versus temperature for N5 alloy.

We think that this area (Fig. 84) of the XRD image is a transmission topograph of dendrites in the solidifying casting. Bright areas in the spot represent reflections from crystalline solid with a [001] orientation. Dark areas are either molten alloy or misaligned areas of the solid.

A topograph is a specular reflection from crystalline planes. Figure 85 shows the geometry of topography. Reflections occur from lattice planes in correctly aligned dendrites. There are no reflections from the amorphous liquid surrounding the dendrites. As can be seen from the figure, there is a vertical exaggeration of the actual structure in the topograph. Additionally, there is a vertical enlargement due to divergence of the collimated x-ray beam. For the geometry of this experiment, this vertical magnification is estimated to be 1.8 (divergence of 1.6 × 1.1 topographic magnification). The horizontal enlargement of structure in the topograph is caused solely by divergence of the x-ray beam, so horizontal magnification for this experiment is approximately 1.6.

Scans in the casting showed that the XRD spot intensity changed, as before, with vertical position in the casting. The topograph became less mottled and more solid when the x-ray beam was directed into the solidified crystal and had a more threadlike appearance higher in the casting where there was more molten alloy and less solid. The XRD spots, including the topograph, disappeared when the x-ray beam passed through only molten N5. The 001 topograph was best displayed during specimen rotations, where the three-dimensional character of the topograph could be appreciated.

88

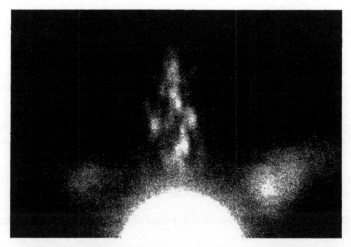

Figure 84. Transmission topograph (top center) of dendrites in an N5 casting which are aligned with the [001] direction. The scale can be inferred by comparison with the diameter of the primary beam (large bright area, whose actual size at the specimen is 3.5 mm). There is an additional vertical exaggeration (1.1) caused by the topographic geometry. The dendrites appear to be approximately 1 to 2.5 mm wide.

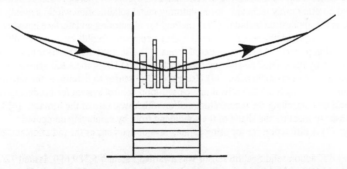

Figure 85. Geometry of transmission topography. X-rays reflected from lattice planes form an image of structure within the casting which have a particular crystalline orientation.

89

6. CONCLUSIONS

1. Transmission x-ray diffraction can serve as a noncontact sensor for locating the liquid–solid boundary while a casting solidifies. Position, velocity, and the shape of the solidification front may be determined.

2. High-contrast transmission diffraction patterns were observable for a single-crystal, nickel-alloy specimen 6 mm thick, surrounded by mold walls 12 mm thick, during casting in a directional solidification furnace.

3. Grain growth in a heated polycrystalline specimen could be monitored by observing changes in the diffraction pattern from a complex array of small diffraction spots to a simple circular pattern containing only a few, high-intensity spots.

4. Defects in crystal growth (bicrystal or changes in crystalline orientation) were conspicuous during casting as changes in the transmission diffraction pattern.

5. The intensity of a diffraction spot was directly related to the solid fraction in the mushy zone of an alloy casting.

6. Precise alignment of the beam and detector to the crystal orientation permits topographic imaging of individual dendrites in the solidfying front.

7. PATENT

NIST filed a patent application for the x-ray diffraction solidification sensing technique on behalf of the Consortium on the Casting of Aerospace Alloys on 24 March 1995. A patent search had disclosed no other x-ray technique for monitoring metal solidification within a mold. The claims made in the application include: (1) A method for monitoring an interface between crystalline and amorphous phases of a material within a container (mold) by using the differences in x-ray diffraction patterns of the two material forms. (2) Using the XRD sensor to define the regions occupied by the crystalline and amorphous phases. (3) Using the XRD sensor to determine the degree of crystallization. (4) Using the XRD sensor to determine the rate of crystallization. (5) Using the XRD sensor in a closed-loop control system for producing castings with high yield by controlling the temperature and/or withdrawal rate of the furnace. (6) Using the XRD sensor to measure the shape of the solidifying front by employing computed tomography. (7) A diffraction sensor using gamma rays or neutrons as the radiation source.

The x-ray diffraction solidification sensor was awarded Patent # 5,589,690; Issued 12/31/96; "Apparatus and Method for Monitoring Casting Process"; Inventors: Thomas A. Siewert, William P. Dubé, and Dale W. Fitting. A copy of the patent is included as an appendix to this report.

The authors are grateful to Steve Seltzer at NIST-Gaithersburg for providing a copy of the XCOM software. They also appreciate the financial support of this measurement technology development project from the NIST Office of Intelligent Processing of Materials and technical interactions with the other members of the NIST Aerospace Casting Consortium.

8. REFERENCES

[1] J.B. Walter and K.L. Telschow, "Laser ultrasonic detection of the solid front during casting," *Review of Progress in Quantitative Nondestructive Evaluation*, D.O. Thompson and D.E. Chimenti (eds.), Vol. 15A, 607-613, Plenum Press, New York, 1996

[2] Y. Tsukahara, C.K. Jen, and C. Neron, "Monitoring the liquid/solid interface with plate waves in the strip casting of metal plates," *Review of Progress in Quantitative Nondestructive Evaluation*, D.O. Thompson and D.E. Chimenti (eds.), Vol. 13B, 2245-2252, Plenum Press, New York, 1994

[3] D.T. Queheillalt, Y. Lou, and H.N.G. Wadley, "Ultrasonic sensing of the location and shape of a solid/liquid interface for crystal growth control," *Review of Progress in Quantitative Nondestructive Evaluation*, D.O. Thompson and D.E. Chimenti (eds.), Vol. 13B, 2253-2260, Plenum Press, New York, 1994

[4] S. Goldspiel, "Radiographic control of castings," *Nondestructive Testing Handbook (2nd edition) on Radiography and Radiation Testing*, L.E. Bryant and P. McIntyre (eds.), Section 10, 458-490, 1985

[5] R.H. Bossi, J.L. Cline, and G.E. Georgeson, "X-ray computed tomographic inspection of castings," *Review of Progress in Quantitative Nondestructive Evaluation*, D.O. Thompson and D.E. Chimenti (eds.), Vol. 10B, 1783-1790, Plenum Press, New York, 1991

[6] J.R. Helliwell, S. Harrop, J. Habash, B.G. Magarian, N.M. Allinnson, D. Gomez, M. Helliwell, Z. Derowenda, and D.W.J. Cruickshank, "Instrumentation for Laue diffraction," *Rev. Sci. Instrum.*, Vol. 60, No. 7, 1531-1536 (1989)

[7] I.J. Clifton, M. Elder, and J. Hajdu, "Experimental strategies in Laue Crystallography," *J. Appl. Crystallogr.*, Vol. 24, 267-277 (1991)

[8] R.E. Green, Jr., "An electro-optical x-ray diffraction system for grain boundary migration measurements at temperature," *Advances in X-Ray Analysis*, Plenum Press (New York), Vol. 15, (1972)

[9] S.E. Doyle, A.R. Gerson, K.J. Roberts, and J.N. Sherwood, "Probing the structure of solids in a liquid environment: a recent in-situ crystallization experiment using high energy wavelength scanning," *J. Cryst. Growth*, Vol. 112, 302-307 (1991)

[10] N.R. Joshi and R.E. Green, Jr., "Continuous x-ray diffraction measurement of lattice rotation during tensile deformation of aluminum crystals," *J. Mater. Sci.*, Vol. 15, 729-738 (1980)

[11] O. Babushkin, R. Harrysson, T. Linback, and R. Tegman, "A high-temperature graphite furnace for x-ray powder diffraction," *Meas. Sci. Tech.*, Vol. 4, 816-819 (1993)

[12] K.G. Abdulvakhidov and M.F. Kupriyanov, "High-temperature unit for x-ray diffraction studies of single crystals," *Instr. Exper. Tech.*, Vol. 35, No. 5, 942-943 (1992)

[13] H.L. Bhat, S.M. Clark, A. Elkorashy, and K.J. Roberts, "A furnace for in-situ synchrotron Laue diffraction and its application to studies of solid-state phase transformations," *J. Appl. Crystallogr.*, Vol 23, 545-549 (1990)

[14] S. Hosokawa, S. Yamada, and K. Tamura, "X-ray diffraction measurements for expanded liquid alkali halides at very high temperature," *J. Non-Cryst. Solids*, Vol. 15, 40-43 (1992)

[15] R.E. Green, Jr., "Applications of flash x-ray diffraction systems to materials testing," *Proc. Flash Rad. Symp., ASNT Fall Conf.*, 151-164 (1977)

[16] H.J. Kopinek, H.H. Otten, and H.J. Bunge, "On-line measuring of technological data of cold and hot rolled steel strips by a fixed angle texture analyer," *Proc. 3rd Int. Symp. on Nondestructive Materials Characterization*, 750-762 (1989)

[17] C.J. Bechtoldt, R.C. Placious, W.J. Boettinger, M. Kuriyama, "X-ray residual stress mapping in industrial materials by energy-dispersive diffractometry," *Advances in X-Ray Analysis*, Plenum Press, New York, Vol. 25, 329-338 (1982)

[18] B.E. Warren, *X-Ray Diffraction*, 381pp, Dover, New York (1990)

[19] L.H. Schwartz and J.B. Cohen, *Diffraction from Materials*, 2nd ed., Springer-Verlag, Berlin (1987)

[20] B.D Cullity, *Elements of X-Ray Diffraction*, 2nd ed., p 555, Addison Wesley (1978)

[21] J.H. Hubbell, H.M. Gerstenberg, and E.B. Saloman, *Bibliography of Photon Total Cross Section (Attenuation Coefficient) Measurements 10 eV to 13.5 GeV*, Nat. Bur. Stand. (U.S.) Internal Report NBSIR 86-3461 (1986).

[22] M.J. Berger and J.H. Hubbell, *XCOM: Photon cross-sections on a personal computer*, Nat. Bur. Stand. (U.S.) Internal Report (NBSIR) 87-3567 (1987). The XCOM software package, written for a personal computer, provides photon cross sections and attenuation coefficients, as a function of photon energy.

[23] L.A. Azaroff, *Elements of X-Ray Crystallography*, Tech Books, Fairfax, VA, p 105 (1968)

[24] B.D. Cullity, *ibid*, pp 156-158

[25] C. T. Chantler, "Theoretical Form Factor, Attenuation, and Scattering Tabulation for Z = 1–92 from E = 1–10 eV to E = 0.4–1.0 MeV," *J. Phys. Chem. Ref. Data*, Vol. 24, 71 (1995)

[26] D. Waasmaier and A. Kirfel, New analytical scattering factor functions for free atoms and ions, *Acta Crystallogr. A*, Vol. 151, No. 3, 416-430 (1995)

9. BIBLIOGRAPHY

The list below contains publications which describe the high-energy transmission technique for monitoring solidification of single-crystal turbine-blade castings.

1. T.A. Siewert, W.P. Dubé, and D.W. Fitting, US Patent 5,589,690, "Apparatus and method for monitoring casting process," issued December 31, 1996

2. D.W. Fitting, W.P. Dubé, and T.A. Siewert, "High energy x-ray diffraction technique for monitoring solidification of single crystal castings," *Proc. Eighth Int. Symp. on Nondestructive Materials Characterization*, Plenum Press, New York, (1998)

3. D.W. Fitting, W.P. Dubé, and T.A. Siewert, "Real-time monitoring of turbine blade solidification using x-ray diffraction techniques," *Semi-Annual and Annual Reports of the NIST Consortium on Casting of Aerospace Alloys*, (1994-1998) (available to Consortuium members)

4. W.P. Dubé, D.W. Fitting, and T.A. Siewert, "Real-Time Sensing of the Liquid-Solid Interface of Castings Using Transmission XRD: Experiments and an Analytical Model," *Denver X-Ray Conf.*, 5-8 August 1996 (Denver, CO)

5. T.A. Siewert, D.W. Fitting, and W.P. Dubé, "Solidification sensing using high-energy x-ray diffraction," *Advanced Materials and Processes*, Vol. 150, No. 1; (1996)

6. D.W. Fitting, W.P. Dubé, T.A. Siewert, and J. Paran, "A process sensor for locating the liquid-solid boundary through the mold of a casting," *Proc. Review of Progress in Quantitative nondestructive Evaluation*, Plenum Press, New York, (1995)

7. D.W. Fitting, W.P. Dubé, and T.A. Siewert, "Real-time sensing of metal solidification using transmission x-ray diffraction," submitted to *Research in Nondestructive Evaluation* (1997)

10. APPENDICES

10.1 Tables of X-Ray Attenuation Coefficients as a Function of Energy

Partial and total mass attenuation coefficients (cm^2/g) were computed with the XCOM software [22]. Data were computed for energies in the 20 to 500 keV range. These are the data we used in the transmission x-ray diffraction modeling.

The partial interaction coefficients are tabulated for the coherent, Compton, and photoelectric processes. The total attenuation coefficients are given with and without the contribution of the coherent process.

Aluminum
Constituents (Átomic Number:Fraction by Weight)
13:1.00000
Partial Interaction Coefficients and Total Attenuation Coefficients

Photon Energy (MeV)	coh (cm2/g)	Comp (cm2/g)	PE (cm2/g)	TOTAL w/coh (cm2/g)	w/o coh (cm2/g)
2.000E-02	2.05E-01	1.37E-01	3.10E+00	3.44E+00	3.24E+00
3.000E-02	1.10E-01	1.46E-01	8.72E-01	1.13E+00	1.02E+00
4.000E-02	6.86E-02	1.49E-01	3.50E-01	5.68E-01	5.00E-01
5.000E-02	4.68E-02	1.50E-01	1.72E-01	3.68E-01	3.21E-01
6.000E-02	3.39E-02	1.48E-01	9.56E-02	2.78E-01	2.44E-01
7.000E-02	2.56E-02	1.46E-01	5.82E-02	2.30E-01	2.04E-01
8.000E-02	2.00E-02	1.44E-01	3.78E-02	2.02E-01	1.82E-01
9.000E-02	1.61E-02	1.41E-01	2.59E-02	1.83E-01	1.67E-01
1.000E-01	1.32E-02	1.39E-01	1.84E-02	1.70E-01	1.57E-01
1.100E-01	1.11E-02	1.36E-01	1.35E-02	1.61E-01	1.50E-01
1.200E-01	9.38E-03	1.34E-01	1.02E-02	1.53E-01	1.44E-01
1.300E-01	8.06E-03	1.31E-01	7.90E-03	1.47E-01	1.39E-01
1.400E-01	6.99E-03	1.29E-01	6.23E-03	1.42E-01	1.35E-01
1.500E-01	6.12E-03	1.27E-01	4.99E-03	1.38E-01	1.32E-01
1.600E-01	5.41E-03	1.25E-01	4.06E-03	1.34E-01	1.29E-01
1.700E-01	4.81E-03	1.23E-01	3.35E-03	1.31E-01	1.26E-01
1.800E-01	4.30E-03	1.21E-01	2.79E-03	1.28E-01	1.23E-01
1.900E-01	3.87E-03	1.19E-01	2.35E-03	1.25E-01	1.21E-01
2.000E-01	3.50E-03	1.17E-01	2.00E-03	1.22E-01	1.19E-01
2.100E-01	3.19E-03	1.15E-01	1.72E-03	1.20E-01	1.17E-01
2.200E-01	2.91E-03	1.13E-01	1.49E-03	1.18E-01	1.15E-01
2.300E-01	2.67E-03	1.12E-01	1.29E-03	1.16E-01	1.13E-01
2.400E-01	2.45E-03	1.10E-01	1.13E-03	1.14E-01	1.11E-01
2.500E-01	2.26E-03	1.09E-01	9.99E-04	1.12E-01	1.10E-01
2.600E-01	2.09E-03	1.07E-01	8.86E-04	1.10E-01	1.08E-01
2.700E-01	1.94E-03	1.06E-01	7.90E-04	1.09E-01	1.07E-01
2.800E-01	1.81E-03	1.05E-01	7.07E-04	1.07E-01	1.05E-01
2.900E-01	1.69E-03	1.03E-01	6.36E-04	1.06E-01	1.04E-01
3.000E-01	1.58E-03	1.02E-01	5.74E-04	1.04E-01	1.03E-01
3.100E-01	1.48E-03	1.01E-01	5.21E-04	1.03E-01	1.01E-01
3.200E-01	1.39E-03	9.97E-02	4.74E-04	1.02E-01	1.00E-01
3.300E-01	1.31E-03	9.86E-02	4.33E-04	1.00E-01	9.90E-02
3.400E-01	1.23E-03	9.75E-02	3.96E-04	9.91E-02	9.79E-02
3.500E-01	1.16E-03	9.64E-02	3.64E-04	9.80E-02	9.68E-02
3.600E-01	1.10E-03	9.54E-02	3.35E-04	9.69E-02	9.58E-02
3.700E-01	1.04E-03	9.44E-02	3.10E-04	9.58E-02	9.47E-02

Aluminum

Photon Energy (MeV)	coh (cm2/g)	Comp (cm2/g)	PE (cm2/g)	TOTAL w/coh (cm2/g)	w/o coh (cm2/g)
3.800E-01	9.89E-04	9.35E-02	2.87E-04	9.47E-02	9.38E-02
3.900E-01	9.40E-04	9.25E-02	2.66E-04	9.37E-02	9.28E-02
4.000E-01	8.93E-04	9.16E-02	2.48E-04	9.28E-02	9.19E-02
4.100E-01	8.51E-04	9.07E-02	2.31E-04	9.18E-02	9.10E-02
4.200E-01	8.11E-04	8.99E-02	2.16E-04	9.09E-02	9.01E-02
4.300E-01	7.74E-04	8.90E-02	2.02E-04	9.00E-02	8.92E-02
4.400E-01	7.39E-04	8.82E-02	1.90E-04	8.92E-02	8.84E-02
4.500E-01	7.07E-04	8.74E-02	1.79E-04	8.83E-02	8.76E-02
4.600E-01	6.77E-04	8.67E-02	1.68E-04	8.75E-02	8.68E-02
4.700E-01	6.48E-04	8.59E-02	1.59E-04	8.67E-02	8.61E-02
4.800E-01	6.22E-04	8.52E-02	1.50E-04	8.59E-02	8.53E-02
4.900E-01	5.97E-04	8.44E-02	1.42E-04	8.52E-02	8.46E-02
5.000E-01	5.73E-04	8.37E-02	1.34E-04	8.45E-02	8.39E-02

Aluminum Oxide
   Constituents (Atomic Number:Fraction by Weight)
   8:0.47075 13:0.52925

Partial Interaction Coefficients and Total Attenuation Coefficients

Photon Energy (MeV)	coh (cm2/g)	Comp (cm2/g)	PE (cm2/g)	TOTAL w/coh (cm2/g)	w/o coh (cm2/g)
2.000E-02	1.55E-01	1.45E-01	1.93E+00	2.23E+00	2.07E+00
3.000E-02	8.27E-02	1.53E-01	5.39E-01	7.75E-01	6.92E-01
4.000E-02	5.14E-02	1.56E-01	2.16E-01	4.23E-01	3.71E-01
5.000E-02	3.50E-02	1.55E-01	1.05E-01	2.95E-01	2.60E-01
6.000E-02	2.53E-02	1.53E-01	5.85E-02	2.37E-01	2.12E-01
7.000E-02	1.91E-02	1.51E-01	3.56E-02	2.05E-01	1.86E-01
8.000E-02	1.49E-02	1.48E-01	2.31E-02	1.86E-01	1.71E-01
9.000E-02	1.20E-02	1.45E-01	1.58E-02	1.73E-01	1.61E-01
1.000E-01	9.82E-03	1.42E-01	1.12E-02	1.63E-01	1.53E-01
1.100E-01	8.20E-03	1.39E-01	8.23E-03	1.56E-01	1.48E-01
1.200E-01	6.95E-03	1.37E-01	6.21E-03	1.50E-01	1.43E-01
1.300E-01	5.96E-03	1.34E-01	4.80E-03	1.45E-01	1.39E-01
1.400E-01	5.17E-03	1.32E-01	3.78E-03	1.41E-01	1.36E-01
1.500E-01	4.53E-03	1.29E-01	3.03E-03	1.37E-01	1.32E-01
1.600E-01	4.00E-03	1.27E-01	2.46E-03	1.34E-01	1.30E-01
1.700E-01	3.55E-03	1.25E-01	2.03E-03	1.31E-01	1.27E-01
1.800E-01	3.18E-03	1.23E-01	1.69E-03	1.28E-01	1.25E-01
1.900E-01	2.86E-03	1.21E-01	1.43E-03	1.25E-01	1.22E-01
2.000E-01	2.59E-03	1.19E-01	1.21E-03	1.23E-01	1.20E-01
2.100E-01	2.35E-03	1.17E-01	1.04E-03	1.21E-01	1.18E-01
2.200E-01	2.15E-03	1.16E-01	8.99E-04	1.19E-01	1.17E-01
2.300E-01	1.97E-03	1.14E-01	7.83E-04	1.17E-01	1.15E-01
2.400E-01	1.81E-03	1.12E-01	6.86E-04	1.15E-01	1.13E-01
2.500E-01	1.67E-03	1.11E-01	6.05E-04	1.13E-01	1.11E-01
2.600E-01	1.54E-03	1.09E-01	5.36E-04	1.11E-01	1.10E-01
2.700E-01	1.43E-03	1.08E-01	4.78E-04	1.10E-01	1.08E-01
2.800E-01	1.33E-03	1.07E-01	4.28E-04	1.08E-01	1.07E-01
2.900E-01	1.24E-03	1.05E-01	3.84E-04	1.07E-01	1.06E-01
3.000E-01	1.16E-03	1.04E-01	3.47E-04	1.06E-01	1.04E-01
3.100E-01	1.09E-03	1.03E-01	3.15E-04	1.04E-01	1.03E-01
3.200E-01	1.02E-03	1.02E-01	2.86E-04	1.03E-01	1.02E-01
3.300E-01	9.64E-04	1.00E-01	2.61E-04	1.02E-01	1.01E-01
3.400E-01	9.08E-04	9.93E-02	2.39E-04	1.00E-01	9.96E-02
3.500E-01	8.57E-04	9.82E-02	2.20E-04	9.93E-02	9.85E-02
3.600E-01	8.11E-04	9.72E-02	2.03E-04	9.82E-02	9.74E-02

Aluminum Oxide

Photon Energy (MeV)	coh (cm2/g)	Comp (cm2/g)	PE (cm2/g)	TOTAL w/coh (cm2/g)	w/o coh (cm2/g)
3.700E-01	7.68E-04	9.62E-02	1.87E-04	9.71E-02	9.64E-02
3.800E-01	7.28E-04	9.52E-02	1.73E-04	9.61E-02	9.54E-02
3.900E-01	6.92E-04	9.42E-02	1.61E-04	9.51E-02	9.44E-02
4.000E-01	6.58E-04	9.33E-02	1.50E-04	9.41E-02	9.35E-02
4.100E-01	6.26E-04	9.24E-02	1.40E-04	9.32E-02	9.26E-02
4.200E-01	5.97E-04	9.15E-02	1.31E-04	9.23E-02	9.17E-02
4.300E-01	5.70E-04	9.07E-02	1.22E-04	9.14E-02	9.08E-02
4.400E-01	5.44E-04	8.99E-02	1.15E-04	9.05E-02	9.00E-02
4.500E-01	5.20E-04	8.90E-02	1.08E-04	8.97E-02	8.92E-02
4.600E-01	4.98E-04	8.83E-02	1.01E-04	8.89E-02	8.84E-02
4.700E-01	4.77E-04	8.75E-02	9.57E-05	8.81E-02	8.76E-02
4.800E-01	4.58E-04	8.67E-02	9.04E-05	8.73E-02	8.68E-02
4.900E-01	4.39E-04	8.60E-02	8.56E-05	8.65E-02	8.61E-02
5.000E-01	4.22E-04	8.53E-02	8.11E-05	8.58E-02	8.54E-02

Borosilicate Glass

Constituents (Atomic Number:Fraction by Weight)
5:0.04010 8:0.53965 11:0.02820 13:0.01170 14:0.37704 19:0.00330

Partial Interaction Coefficients and Total Attenuation Coefficients

Photon				TOTAL	
Energy	coh	Comp	PE	w/coh	w/o coh
(MeV)	(cm2/g)	(cm2/g)	(cm2/g)	(cm2/g)	(cm2/g)
2.000E-02	1.52E-01	1.48E-01	2.00E+00	2.30E+00	2.14E+00
3.000E-02	8.09E-02	1.56E-01	5.61E-01	7.99E-01	7.18E-01
4.000E-02	5.04E-02	1.58E-01	2.25E-01	4.34E-01	3.84E-01
5.000E-02	3.43E-02	1.57E-01	1.11E-01	3.02E-01	2.68E-01
6.000E-02	2.48E-02	1.55E-01	6.15E-02	2.42E-01	2.17E-01
7.000E-02	1.87E-02	1.53E-01	3.75E-02	2.09E-01	1.90E-01
8.000E-02	1.46E-02	1.50E-01	2.44E-02	1.89E-01	1.74E-01
9.000E-02	1.18E-02	1.47E-01	1.66E-02	1.76E-01	1.64E-01
1.000E-01	9.65E-03	1.44E-01	1.18E-02	1.66E-01	1.56E-01
1.100E-01	8.06E-03	1.41E-01	8.71E-03	1.58E-01	1.50E-01
1.200E-01	6.83E-03	1.39E-01	6.58E-03	1.52E-01	1.45E-01
1.300E-01	5.86E-03	1.36E-01	5.09E-03	1.47E-01	1.41E-01
1.400E-01	5.08E-03	1.34E-01	4.01E-03	1.43E-01	1.38E-01
1.500E-01	4.45E-03	1.31E-01	3.21E-03	1.39E-01	1.34E-01
1.600E-01	3.93E-03	1.29E-01	2.62E-03	1.36E-01	1.32E-01
1.700E-01	3.49E-03	1.27E-01	2.16E-03	1.32E-01	1.29E-01
1.800E-01	3.12E-03	1.25E-01	1.80E-03	1.30E-01	1.27E-01
1.900E-01	2.81E-03	1.23E-01	1.52E-03	1.27E-01	1.24E-01
2.000E-01	2.54E-03	1.21E-01	1.29E-03	1.25E-01	1.22E-01
2.100E-01	2.31E-03	1.19E-01	1.11E-03	1.22E-01	1.20E-01
2.200E-01	2.11E-03	1.17E-01	9.57E-04	1.20E-01	1.18E-01
2.300E-01	1.93E-03	1.16E-01	8.33E-04	1.18E-01	1.16E-01
2.400E-01	1.78E-03	1.14E-01	7.30E-04	1.16E-01	1.15E-01
2.500E-01	1.64E-03	1.12E-01	6.44E-04	1.15E-01	1.13E-01
2.600E-01	1.52E-03	1.11E-01	5.71E-04	1.13E-01	1.11E-01
2.700E-01	1.41E-03	1.09E-01	5.09E-04	1.11E-01	1.10E-01
2.800E-01	1.31E-03	1.08E-01	4.56E-04	1.10E-01	1.09E-01
2.900E-01	1.22E-03	1.07E-01	4.10E-04	1.08E-01	1.07E-01
3.000E-01	1.14E-03	1.05E-01	3.70E-04	1.07E-01	1.06E-01
3.100E-01	1.07E-03	1.04E-01	3.36E-04	1.06E-01	1.05E-01
3.200E-01	1.01E-03	1.03E-01	3.06E-04	1.04E-01	1.03E-01
3.300E-01	9.48E-04	1.02E-01	2.79E-04	1.03E-01	1.02E-01
3.400E-01	8.93E-04	1.01E-01	2.56E-04	1.02E-01	1.01E-01
3.500E-01	8.43E-04	9.96E-02	2.35E-04	1.01E-01	9.98E-02

Borosilicate Glass

Photon Energy (MeV)	coh (cm2/g)	Comp (cm2/g)	PE (cm2/g)	TOTAL w/coh (cm2/g)	w/o coh (cm2/g)
3.600E-01	7.98E-04	9.85E-02	2.16E-04	9.95E-02	9.87E-02
3.700E-01	7.55E-04	9.75E-02	2.00E-04	9.85E-02	9.77E-02
3.800E-01	7.16E-04	9.65E-02	1.85E-04	9.74E-02	9.67E-02
3.900E-01	6.80E-04	9.55E-02	1.72E-04	9.64E-02	9.57E-02
4.000E-01	6.47E-04	9.46E-02	1.60E-04	9.54E-02	9.48E-02
4.100E-01	6.16E-04	9.37E-02	1.49E-04	9.44E-02	9.38E-02
4.200E-01	5.87E-04	9.28E-02	1.39E-04	9.35E-02	9.29E-02
4.300E-01	5.60E-04	9.19E-02	1.31E-04	9.26E-02	9.21E-02
4.400E-01	5.35E-04	9.11E-02	1.23E-04	9.17E-02	9.12E-02
4.500E-01	5.12E-04	9.03E-02	1.15E-04	9.09E-02	9.04E-02
4.600E-01	4.90E-04	8.95E-02	1.08E-04	9.01E-02	8.96E-02
4.700E-01	4.69E-04	8.87E-02	1.02E-04	8.93E-02	8.88E-02
4.800E-01	4.50E-04	8.79E-02	9.67E-05	8.85E-02	8.80E-02
4.900E-01	4.32E-04	8.72E-02	9.15E-05	8.77E-02	8.73E-02
5.000E-01	4.15E-04	8.65E-02	8.67E-05	8.70E-02	8.65E-02

Copper
Constituents (Atomic Number:Fraction by Weight)
29:1.00000

Partial Interaction Coefficients and Total Attenuation Coefficients

Photon Energy (MeV)	coh (cm2/g)	Comp (cm2/g)	PE (cm2/g)	TOTAL	
				w/coh (cm2/g)	w/o coh (cm2/g)
2.000E-02	6.06E-01	1.10E-01	3.31E+01	3.38E+01	3.32E+01
3.000E-02	3.37E-01	1.23E-01	1.05E+01	1.09E+01	1.06E+01
4.000E-02	2.12E-01	1.29E-01	4.52E+01	4.86E+00	4.65E+00
5.000E-02	1.47E-01	1.31E-01	2.34E+00	2.61E+00	2.47E+00
6.000E-02	1.08E-01	1.31E-01	1.35E+00	1.59E+00	1.48E+00
7.000E-02	8.30E-02	1.31E-01	8.50E-01	1.06E+00	9.81E-01
8.000E-02	6.59E-02	1.29E-01	5.68E-01	7.63E-01	6.97E-01
9.000E-02	5.36E-02	1.28E-01	3.97E-01	5.78E-01	5.25E-01
1.000E-01	4.45E-02	1.26E-01	2.88E-01	4.58E-01	4.14E-01
1.100E-01	3.74E-02	1.24E-01	2.15E-01	3.77E-01	3.40E-01
1.200E-01	3.19E-02	1.23E-01	1.65E-01	3.19E-01	2.88E-01
1.300E-01	2.76E-02	1.21E-01	1.29E-01	2.77E-01	2.50E-01
1.400E-01	2.40E-02	1.19E-01	1.03E-01	2.46E-01	2.22E-01
1.500E-01	2.11E-02	1.17E-01	8.35E-02	2.22E-01	2.01E-01
1.600E-01	1.87E-02	1.15E-01	6.86E-02	2.03E-01	1.84E-01
1.700E-01	1.67E-02	1.14E-01	5.70E-02	1.87E-01	1.71E-01
1.800E-01	1.50E-02	1.12E-01	4.79E-02	1.75E-01	1.60E-01
1.900E-01	1.35E-02	1.10E-01	4.07E-02	1.65E-01	1.51E-01
2.000E-01	1.23E-02	1.09E-01	3.49E-02	1.56E-01	1.44E-01
2.100E-01	1.12E-02	1.07E-01	3.01E-02	1.49E-01	1.37E-01
2.200E-01	1.02E-02	1.06E-01	2.62E-02	1.42E-01	1.32E-01
2.300E-01	9.39E-03	1.04E-01	2.29E-02	1.37E-01	1.27E-01
2.400E-01	8.65E-03	1.03E-01	2.02E-02	1.32E-01	1.23E-01
2.500E-01	8.00E-03	1.02E-01	1.79E-02	1.28E-01	1.20E-01
2.600E-01	7.41E-03	1.00E-01	1.60E-02	1.24E-01	1.16E-01
2.700E-01	6.89E-03	9.93E-02	1.43E-02	1.20E-01	1.14E-01
2.800E-01	6.42E-03	9.81E-02	1.28E-02	1.17E-01	1.11E-01
2.900E-01	6.00E-03	9.69E-02	1.16E-02	1.15E-01	1.09E-01
3.000E-01	5.62E-03	9.58E-02	1.05E-02	1.12E-01	1.06E-01
3.100E-01	5.27E-03	9.47E-02	9.56E-03	1.10E-01	1.04E-01
3.200E-01	4.96E-03	9.37E-02	8.73E-03	1.07E-01	1.02E-01
3.300E-01	4.67E-03	9.26E-02	7.99E-03	1.05E-01	1.01E-01
3.400E-01	4.41E-03	9.17E-02	7.34E-03	1.03E-01	9.90E-02
3.500E-01	4.16E-03	9.07E-02	6.77E-03	1.02E-01	9.75E-02
3.600E-01	3.94E-03	8.97E-02	6.25E-03	9.99E-02	9.60E-02

Copper

Photon Energy (MeV)	coh (cm2/g)	Comp (cm2/g)	PE (cm2/g)	TOTAL w/coh (cm2/g)	w/o coh (cm2/g)
3.700E-01	3.74E-03	8.88E-02	5.79E-03	9.84E-02	9.46E-02
3.800E-01	3.55E-03	8.80E-02	5.37E-03	9.69E-02	9.33E-02
3.900E-01	3.37E-03	8.71E-02	5.00E-03	9.55E-02	9.21E-02
4.000E-01	3.21E-03	8.63E-02	4.66E-03	9.41E-02	9.09E-02
4.100E-01	3.06E-03	8.54E-02	4.36E-03	9.29E-02	8.98E-02
4.200E-01	2.91E-03	8.47E-02	4.08E-03	9.16E-02	8.87E-02
4.300E-01	2.78E-03	8.39E-02	3.83E-03	9.05E-02	8.77E-02
4.400E-01	2.66E-03	8.31E-02	3.60E-03	8.94E-02	8.67E-02
4.500E-01	2.54E-03	8.24E-02	3.39E-03	8.83E-02	8.58E-02
4.600E-01	2.44E-03	8.17E-02	3.20E-03	8.73E-02	8.49E-02
4.700E-01	2.34E-03	8.10E-02	3.02E-03	8.63E-02	8.40E-02
4.800E-01	2.24E-03	8.03E-02	2.86E-03	8.54E-02	8.32E-02
4.900E-01	2.15E-03	7.96E-02	2.71E-03	8.45E-02	8.23E-02
5.000E-01	2.07E-03	7.90E-02	2.57E-03	8.36E-02	8.16E-02

Gadolinium Oxysulfide
   Constituents (Atomic Number:Fraction by Weight)
      8:0.08453 16:0.08470 64:0.83077

Partial Interaction Coefficients and Total Attenuation Coefficients

Photon Energy (MeV)	coh (cm2/g)	Comp (cm2/g)	PE (cm2/g)	TOTAL w/coh (cm2/g)	TOTAL w/o coh (cm2/g)
2.000E-02	1.39E+00	9.03E-02	3.54E+01	3.69E+01	3.55E+01
3.000E-02	8.18E-01	1.03E-01	1.16E+01	1.25E+01	1.17E+01
4.000E-02	5.33E-01	1.09E-01	5.21E+00	5.85E+00	5.32E+00
5.000E-02	3.77E-01	1.12E-01	2.78E+00	3.27E+00	2.90E+00
6.000E-02	2.83E-01	1.13E-01	9.42E+00	9.81E+00	9.53E+00
7.000E-02	2.21E-01	1.14E-01	6.26E+00	6.59E+00	6.37E+00
8.000E-02	1.77E-01	1.14E-01	4.38E+00	4.67E+00	4.49E+00
9.000E-02	1.45E-01	1.13E-01	3.18E+00	3.44E+00	3.29E+00
1.000E-01	1.20E-01	1.12E-01	2.38E+00	2.61E+00	2.49E+00
1.100E-01	1.02E-01	1.11E-01	1.83E+00	2.04E+00	1.94E+00
1.200E-01	8.73E-02	1.09E-01	1.44E+00	1.64E+00	1.55E+00
1.300E-01	7.57E-02	1.08E-01	1.15E+00	1.34E+00	1.26E+00
1.400E-01	6.63E-02	1.06E-01	9.38E-01	1.11E+00	1.04E+00
1.500E-01	5.86E-02	1.05E-01	7.74E-01	9.38E-01	8.80E-01
1.600E-01	5.22E-02	1.04E-01	6.47E-01	8.03E-01	7.51E-01
1.700E-01	4.68E-02	1.02E-01	5.47E-01	6.96E-01	6.49E-01
1.800E-01	4.22E-02	1.01E-01	4.66E-01	6.10E-01	5.67E-01
1.900E-01	3.83E-02	9.96E-02	4.01E-01	5.39E-01	5.01E-01
2.000E-01	3.49E-02	9.83E-02	3.48E-01	4.81E-01	4.46E-01
2.100E-01	3.19E-02	9.71E-02	3.04E-01	4.33E-01	4.01E-01
2.200E-01	2.93E-02	9.59E-02	2.67E-01	3.93E-01	3.63E-01
2.300E-01	2.70E-02	9.47E-02	2.37E-01	3.58E-01	3.31E-01
2.400E-01	2.49E-02	9.35E-02	2.10E-01	3.29E-01	3.04E-01
2.500E-01	2.31E-02	9.24E-02	1.88E-01	3.04E-01	2.81E-01
2.600E-01	2.15E-02	9.13E-02	1.69E-01	2.82E-01	2.60E-01
2.700E-01	2.00E-02	9.03E-02	1.53E-01	2.63E-01	2.43E-01
2.800E-01	1.87E-02	8.93E-02	1.38E-01	2.46E-01	2.28E-01
2.900E-01	1.75E-02	8.83E-02	1.26E-01	2.32E-01	2.14E-01
3.000E-01	1.64E-02	8.73E-02	1.15E-01	2.18E-01	2.02E-01
3.100E-01	1.54E-02	8.64E-02	1.05E-01	2.07E-01	1.91E-01
3.200E-01	1.45E-02	8.55E-02	9.66E-02	1.97E-01	1.82E-01
3.300E-01	1.37E-02	8.46E-02	8.90E-02	1.87E-01	1.74E-01
3.400E-01	1.29E-02	8.37E-02	8.22E-02	1.79E-01	1.66E-01
3.500E-01	1.23E-02	8.29E-02	7.62E-02	1.71E-01	1.59E-01
3.600E-01	1.16E-02	8.21E-02	7.08E-02	1.64E-01	1.53E-01

Gadolinium Oxysulfide

Photon Energy (MeV)	coh (cm2/g)	Comp (cm2/g)	PE (cm2/g)	TOTAL w/coh (cm2/g)	w/o coh (cm2/g)
3.700E-01	1.10E-02	8.13E-02	6.59E-02	1.58E-01	1.47E-01
3.800E-01	1.05E-02	8.05E-02	6.15E-02	1.52E-01	1.42E-01
3.900E-01	9.97E-03	7.97E-02	5.75E-02	1.47E-01	1.37E-01
4.000E-01	9.49E-03	7.90E-02	5.39E-02	1.42E-01	1.33E-01
4.100E-01	9.05E-03	7.83E-02	5.05E-02	1.38E-01	1.29E-01
4.200E-01	8.64E-03	7.76E-02	4.75E-02	1.34E-01	1.25E-01
4.300E-01	8.26E-03	7.69E-02	4.48E-02	1.30E-01	1.22E-01
4.400E-01	7.90E-03	7.63E-02	4.22E-02	1.26E-01	1.18E-01
4.500E-01	7.57E-03	7.56E-02	3.99E-02	1.23E-01	1.16E-01
4.600E-01	7.25E-03	7.50E-02	3.78E-02	1.20E-01	1.13E-01
4.700E-01	6.96E-03	7.43E-02	3.58E-02	1.17E-01	1.10E-01
4.800E-01	6.68E-03	7.37E-02	3.40E-02	1.14E-01	1.08E-01
4.900E-01	6.42E-03	7.31E-02	3.23E-02	1.12E-01	1.05E-01
5.000E-01	6.18E-03	7.26E-02	3.07E-02	1.09E-01	1.03E-01

Gallium
  Constituents (Atomic Number:Fraction by Weight)
   31:1.00000

Partial Interaction Coefficients and Total Attenuation Coefficients

Photon				TOTAL	
Energy	coh	Comp	PE	w/coh	w/o coh
(MeV)	(cm2/g)	(cm2/g)	(cm2/g)	(cm2/g)	(cm2/g)
2.000E-02	6.52E-01	1.05E-01	3.85E+01	3.93E+01	3.86E+01
3.000E-02	3.65E-01	1.18E-01	1.23E+01	1.28E+01	1.24E+01
4.000E-02	2.31E-01	1.24E-01	5.37E+00	5.73E+00	5.49E+00
5.000E-02	1.59E-01	1.27E-01	2.79E+00	3.08E+00	2.92E+00
6.000E-02	1.17E-01	1.27E-01	1.62E+00	1.87E+00	1.75E+00
7.000E-02	9.01E-02	1.26E-01	1.02E+00	1.24E+00	1.15E+00
8.000E-02	7.16E-02	1.25E-01	6.85E-01	8.82E-01	8.11E-01
9.000E-02	5.83E-02	1.24E-01	4.80E-01	6.62E-01	6.04E-01
1.000E-01	4.83E-02	1.22E-01	3.49E-01	5.20E-01	4.71E-01
1.100E-01	4.07E-02	1.21E-01	2.61E-01	4.23E-01	3.82E-01
1.200E-01	3.48E-02	1.19E-01	2.01E-01	3.54E-01	3.20E-01
1.300E-01	3.00E-02	1.17E-01	1.57E-01	3.05E-01	2.75E-01
1.400E-01	2.62E-02	1.15E-01	1.26E-01	2.67E-01	2.41E-01
1.500E-01	2.30E-02	1.14E-01	1.02E-01	2.39E-01	2.16E-01
1.600E-01	2.04E-02	1.12E-01	8.38E-02	2.16E-01	1.96E-01
1.700E-01	1.82E-02	1.10E-01	6.98E-02	1.98E-01	1.80E-01
1.800E-01	1.63E-02	1.09E-01	5.87E-02	1.84E-01	1.68E-01
1.900E-01	1.48E-02	1.07E-01	4.99E-02	1.72E-01	1.57E-01
2.000E-01	1.34E-02	1.06E-01	4.28E-02	1.62E-01	1.49E-01
2.100E-01	1.22E-02	1.04E-01	3.70E-02	1.53E-01	1.41E-01
2.200E-01	1.12E-02	1.03E-01	3.22E-02	1.46E-01	1.35E-01
2.300E-01	1.03E-02	1.02E-01	2.82E-02	1.40E-01	1.30E-01
2.400E-01	9.45E-03	1.00E-01	2.49E-02	1.35E-01	1.25E-01
2.500E-01	8.74E-03	9.89E-02	2.21E-02	1.30E-01	1.21E-01
2.600E-01	8.10E-03	9.77E-02	1.97E-02	1.25E-01	1.17E-01
2.700E-01	7.53E-03	9.65E-02	1.76E-02	1.22E-01	1.14E-01
2.800E-01	7.02E-03	9.54E-02	1.58E-02	1.18E-01	1.11E-01
2.900E-01	6.56E-03	9.43E-02	1.43E-02	1.15E-01	1.09E-01
3.000E-01	6.15E-03	9.32E-02	1.30E-02	1.12E-01	1.06E-01
3.100E-01	5.77E-03	9.21E-02	1.18E-02	1.10E-01	1.04E-01
3.200E-01	5.42E-03	9.11E-02	1.08E-02	1.07E-01	1.02E-01
3.300E-01	5.11E-03	9.01E-02	9.87E-03	1.05E-01	1.00E-01
3.400E-01	4.82E-03	8.92E-02	9.07E-03	1.03E-01	9.83E-02
3.500E-01	4.55E-03	8.82E-02	8.36E-03	1.01E-01	9.66E-02
3.600E-01	4.31E-03	8.73E-02	7.73E-03	9.94E-02	9.51E-02

Gallium

Photon Energy (MeV)	coh (cm2/g)	Comp (cm2/g)	PE (cm2/g)	TOTAL w/coh (cm2/g)	w/o coh (cm2/g)
3.700E-01	4.09E-03	8.65E-02	7.16E-03	9.77E-02	9.36E-02
3.800E-01	3.88E-03	8.56E-02	6.65E-03	9.61E-02	9.23E-02
3.900E-01	3.69E-03	8.48E-02	6.19E-03	9.47E-02	9.10E-02
4.000E-01	3.51E-03	8.40E-02	5.77E-03	9.32E-02	8.97E-02
4.100E-01	3.34E-03	8.32E-02	5.40E-03	9.19E-02	8.86E-02
4.200E-01	3.19E-03	8.24E-02	5.06E-03	9.07E-02	8.75E-02
4.300E-01	3.05E-03	8.17E-02	4.75E-03	8.94E-02	8.64E-02
4.400E-01	2.91E-03	8.09E-02	4.46E-03	8.83E-02	8.54E-02
4.500E-01	2.79E-03	8.02E-02	4.20E-03	8.72E-02	8.44E-02
4.600E-01	2.67E-03	7.95E-02	3.96E-03	8.62E-02	8.35E-02
4.700E-01	2.56E-03	7.88E-02	3.74E-03	8.51E-02	8.26E-02
4.800E-01	2.45E-03	7.82E-02	3.54E-03	8.42E-02	8.17E-02
4.900E-01	2.36E-03	7.75E-02	3.36E-03	8.33E-02	8.09E-02
5.000E-01	2.26E-03	7.69E-02	3.19E-03	8.24E-02	8.01E-02

Germanium
Constituents (Atomic Number:Fraction by Weight)
32:1.00000

Partial Interaction Coefficients and Total Attenuation Coefficients

Photon Energy (MeV)	coh (cm2/g)	Comp (cm2/g)	PE (cm2/g)	TOTAL w/coh (cm2/g)	w/o coh (cm2/g)
2.000E-02	6.77E-01	1.03E-01	4.14E+01	4.22E+01	4.15E+01
3.000E-02	3.80E-01	1.16E-01	1.34E+01	1.38E+01	1.35E+01
4.000E-02	2.41E-01	1.22E-01	5.84E+00	6.21E+00	5.97E+00
5.000E-02	1.66E-01	1.25E-01	3.04E+00	3.34E+00	3.17E+00
6.000E-02	1.22E-01	1.25E-01	1.77E+00	2.02E+00	1.90E+00
7.000E-02	9.40E-02	1.25E-01	1.12E+00	1.34E+00	1.25E+00
8.000E-02	7.47E-02	1.24E-01	7.52E-01	9.50E-01	8.75E-01
9.000E-02	6.08E-02	1.23E-01	5.27E-01	7.10E-01	6.50E-01
1.000E-01	5.05E-02	1.21E-01	3.84E-01	5.55E-01	5.05E-01
1.100E-01	4.25E-02	1.19E-01	2.87E-01	4.49E-01	4.07E-01
1.200E-01	3.63E-02	1.18E-01	2.21E-01	3.75E-01	3.39E-01
1.300E-01	3.14E-02	1.16E-01	1.73E-01	3.21E-01	2.89E-01
1.400E-01	2.74E-02	1.14E-01	1.39E-01	2.80E-01	2.53E-01
1.500E-01	2.41E-02	1.13E-01	1.12E-01	2.49E-01	2.25E-01
1.600E-01	2.13E-02	1.11E-01	9.25E-02	2.25E-01	2.03E-01
1.700E-01	1.90E-02	1.09E-01	7.71E-02	2.05E-01	1.86E-01
1.800E-01	1.71E-02	1.08E-01	6.49E-02	1.90E-01	1.73E-01
1.900E-01	1.54E-02	1.06E-01	5.52E-02	1.77E-01	1.61E-01
2.000E-01	1.40E-02	1.05E-01	4.73E-02	1.66E-01	1.52E-01
2.100E-01	1.28E-02	1.03E-01	4.09E-02	1.57E-01	1.44E-01
2.200E-01	1.17E-02	1.02E-01	3.56E-02	1.49E-01	1.38E-01
2.300E-01	1.07E-02	1.01E-01	3.12E-02	1.43E-01	1.32E-01
2.400E-01	9.89E-03	9.93E-02	2.75E-02	1.37E-01	1.27E-01
2.500E-01	9.14E-03	9.80E-02	2.44E-02	1.32E-01	1.22E-01
2.600E-01	8.48E-03	9.68E-02	2.18E-02	1.27E-01	1.19E-01
2.700E-01	7.88E-03	9.56E-02	1.95E-02	1.23E-01	1.15E-01
2.800E-01	7.35E-03	9.45E-02	1.75E-02	1.19E-01	1.12E-01
2.900E-01	6.87E-03	9.34E-02	1.58E-02	1.16E-01	1.09E-01
3.000E-01	6.43E-03	9.23E-02	1.44E-02	1.13E-01	1.07E-01
3.100E-01	6.04E-03	9.13E-02	1.31E-02	1.10E-01	1.04E-01
3.200E-01	5.68E-03	9.03E-02	1.20E-02	1.08E-01	1.02E-01
3.300E-01	5.35E-03	8.93E-02	1.10E-02	1.06E-01	1.00E-01
3.400E-01	5.05E-03	8.83E-02	1.01E-02	1.03E-01	9.84E-02
3.500E-01	4.77E-03	8.74E-02	9.28E-03	1.01E-01	9.67E-02
3.600E-01	4.51E-03	8.65E-02	8.58E-03	9.96E-02	9.51E-02

Germanium

Photon Energy (MeV)	coh (cm2/g)	Comp (cm2/g)	PE (cm2/g)	TOTAL w/coh (cm2/g)	w/o coh (cm2/g)
3.700E-01	4.28E-03	8.57E-02	7.95E-03	9.79E-02	9.36E-02
3.800E-01	4.06E-03	8.48E-02	7.38E-03	9.63E-02	9.22E-02
3.900E-01	3.86E-03	8.40E-02	6.87E-03	9.47E-02	9.09E-02
4.000E-01	3.67E-03	8.32E-02	6.41E-03	9.33E-02	8.96E-02
4.100E-01	3.50E-03	8.24E-02	5.99E-03	9.19E-02	8.84E-02
4.200E-01	3.34E-03	8.16E-02	5.62E-03	9.06E-02	8.73E-02
4.300E-01	3.19E-03	8.09E-02	5.27E-03	8.94E-02	8.62E-02
4.400E-01	3.05E-03	8.02E-02	4.96E-03	8.82E-02	8.51E-02
4.500E-01	2.92E-03	7.95E-02	4.67E-03	8.71E-02	8.41E-02
4.600E-01	2.79E-03	7.88E-02	4.40E-03	8.60E-02	8.32E-02
4.700E-01	2.68E-03	7.81E-02	4.16E-03	8.50E-02	8.23E-02
4.800E-01	2.57E-03	7.75E-02	3.94E-03	8.40E-02	8.14E-02
4.900E-01	2.47E-03	7.68E-02	3.73E-03	8.30E-02	8.06E-02
5.000E-01	2.37E-03	7.62E-02	3.54E-03	8.21E-02	7.97E-02

Mold Material (alumina, silica, and zirconia mixture)
Constituents (Atomic Number:Fraction by Weight)
8:0.42080 13:0.27127 14:0.09067 40:0.21726

Partial Interaction Coefficients and Total Attenuation Coefficients

Photon Energy (MeV)	coh (cm2/g)	Comp (cm2/g)	PE (cm2/g)	TOTAL w/coh (cm2/g)	w/o coh (cm2/g)
2.000E-02	3.19E-01	1.36E-01	1.70E+01	1.74E+01	1.71E+01
3.000E-02	1.76E-01	1.45E-01	5.67E+00	5.99E+00	5.82E+00
4.000E-02	1.13E-01	1.48E-01	2.54E+00	2.80E+00	2.69E+00
5.000E-02	7.79E-02	1.48E-01	1.34E+00	1.57E+00	1.49E+00
6.000E-02	5.70E-02	1.47E-01	7.94E-01	9.98E-01	9.41E-01
7.000E-02	4.35E-02	1.45E-01	5.07E-01	6.95E-01	6.52E-01
8.000E-02	3.44E-02	1.43E-01	3.42E-01	5.19E-01	4.85E-01
9.000E-02	2.79E-02	1.40E-01	2.42E-01	4.10E-01	3.82E-01
1.000E-01	2.31E-02	1.38E-01	1.77E-01	3.38E-01	3.15E-01
1.100E-01	1.94E-02	1.35E-01	1.33E-01	2.88E-01	2.69E-01
1.200E-01	1.66E-02	1.33E-01	1.03E-01	2.53E-01	2.36E-01
1.300E-01	1.43E-02	1.31E-01	8.13E-02	2.26E-01	2.12E-01
1.400E-01	1.25E-02	1.28E-01	6.52E-02	2.06E-01	1.93E-01
1.500E-01	1.10E-02	1.26E-01	5.31E-02	1.90E-01	1.79E-01
1.600E-01	9.72E-03	1.24E-01	4.39E-02	1.78E-01	1.68E-01
1.700E-01	8.67E-03	1.22E-01	3.66E-02	1.67E-01	1.59E-01
1.800E-01	7.78E-03	1.20E-01	3.09E-02	1.59E-01	1.51E-01
1.900E-01	7.02E-03	1.18E-01	2.64E-02	1.52E-01	1.45E-01
2.000E-01	6.37E-03	1.16E-01	2.27E-02	1.45E-01	1.39E-01
2.100E-01	5.80E-03	1.15E-01	1.97E-02	1.40E-01	1.34E-01
2.200E-01	5.31E-03	1.13E-01	1.72E-02	1.35E-01	1.30E-01
2.300E-01	4.88E-03	1.11E-01	1.51E-02	1.31E-01	1.26E-01
2.400E-01	4.49E-03	1.10E-01	1.33E-02	1.28E-01	1.23E-01
2.500E-01	4.15E-03	1.08E-01	1.18E-02	1.24E-01	1.20E-01
2.600E-01	3.85E-03	1.07E-01	1.06E-02	1.21E-01	1.18E-01
2.700E-01	3.58E-03	1.06E-01	9.48E-03	1.19E-01	1.15E-01
2.800E-01	3.34E-03	1.04E-01	8.54E-03	1.16E-01	1.13E-01
2.900E-01	3.12E-03	1.03E-01	7.73E-03	1.14E-01	1.11E-01
3.000E-01	2.92E-03	1.02E-01	7.02E-03	1.12E-01	1.09E-01
3.100E-01	2.74E-03	1.01E-01	6.40E-03	1.10E-01	1.07E-01
3.200E-01	2.57E-03	9.94E-02	5.85E-03	1.08E-01	1.05E-01
3.300E-01	2.42E-03	9.83E-02	5.37E-03	1.06E-01	1.04E-01
3.400E-01	2.29E-03	9.73E-02	4.94E-03	1.04E-01	1.02E-01
3.500E-01	2.16E-03	9.62E-02	4.56E-03	1.03E-01	1.01E-01
3.600E-01	2.05E-03	9.52E-02	4.22E-03	1.01E-01	9.94E-02

Mold Material (alumina, silica, and zirconia mixture)

Photon Energy (MeV)	coh (cm2/g)	Comp (cm2/g)	PE (cm2/g)	TOTAL	
				w/coh (cm2/g)	w/o coh (cm2/g)
3.700E-01	1.94E-03	9.42E-02	3.91E-03	1.00E-01	9.81E-02
3.800E-01	1.84E-03	9.33E-02	3.64E-03	9.87E-02	9.69E-02
3.900E-01	1.75E-03	9.23E-02	3.39E-03	9.75E-02	9.57E-02
4.000E-01	1.67E-03	9.14E-02	3.16E-03	9.63E-02	9.46E-02
4.100E-01	1.59E-03	9.06E-02	2.96E-03	9.51E-02	9.35E-02
4.200E-01	1.51E-03	8.97E-02	2.78E-03	9.40E-02	9.25E-02
4.300E-01	1.45E-03	8.89E-02	2.61E-03	9.29E-02	9.15E-02
4.400E-01	1.38E-03	8.81E-02	2.45E-03	9.19E-02	9.05E-02
4.500E-01	1.32E-03	8.73E-02	2.31E-03	9.09E-02	8.96E-02
4.600E-01	1.27E-03	8.65E-02	2.18E-03	9.00E-02	8.87E-02
4.700E-01	1.21E-03	8.58E-02	2.06E-03	8.90E-02	8.78E-02
4.800E-01	1.16E-03	8.50E-02	1.95E-03	8.82E-02	8.70E-02
4.900E-01	1.12E-03	8.43E-02	1.85E-03	8.73E-02	8.62E-02
5.000E-01	1.07E-03	8.36E-02	1.76E-03	8.65E-02	8.54E-02

Molybdenum
  Constituents (Atomic Number:Fraction by Weight)
  42:1.00000

Partial Interaction Coefficients and Total Attenuation Coefficients

Photon Energy (MeV)	coh (cm2/g)	Comp (cm2/g)	PE (cm2/g)	TOTAL w/coh (cm2/g)	w/o coh (cm2/g)
2.000E-02	9.89E-01	9.51E-02	7.85E+01	7.95E+01	7.86E+01
3.000E-02	5.58E-01	1.09E-01	2.74E+01	2.81E+01	2.75E+01
4.000E-02	3.64E-01	1.16E-01	1.25E+01	1.29E+01	1.26E+01
5.000E-02	2.54E-01	1.19E-01	6.67E+00	7.04E+00	6.79E+00
6.000E-02	1.87E-01	1.20E-01	3.97E+00	4.27E+00	4.09E+00
7.000E-02	1.44E-01	1.21E-01	2.55E+00	2.81E+00	2.67E+00
8.000E-02	1.14E-01	1.20E-01	1.73E+00	1.96E+00	1.85E+00
9.000E-02	9.30E-02	1.19E-01	1.23E+00	1.44E+00	1.34E+00
1.000E-01	7.73E-02	1.18E-01	9.01E-01	1.10E+00	1.02E+00
1.100E-01	6.53E-02	1.16E-01	6.81E-01	8.63E-01	7.97E-01
1.200E-01	5.60E-02	1.15E-01	5.27E-01	6.98E-01	6.42E-01
1.300E-01	4.85E-02	1.13E-01	4.16E-01	5.78E-01	5.30E-01
1.400E-01	4.24E-02	1.12E-01	3.35E-01	4.89E-01	4.46E-01
1.500E-01	3.74E-02	1.10E-01	2.73E-01	4.21E-01	3.83E-01
1.600E-01	3.32E-02	1.09E-01	2.26E-01	3.68E-01	3.35E-01
1.700E-01	2.97E-02	1.07E-01	1.89E-01	3.26E-01	2.96E-01
1.800E-01	2.67E-02	1.06E-01	1.60E-01	2.92E-01	2.66E-01
1.900E-01	2.41E-02	1.04E-01	1.37E-01	2.65E-01	2.41E-01
2.000E-01	2.19E-02	1.03E-01	1.18E-01	2.42E-01	2.20E-01
2.100E-01	2.00E-02	1.02E-01	1.02E-01	2.23E-01	2.03E-01
2.200E-01	1.83E-02	1.00E-01	8.90E-02	2.08E-01	1.89E-01
2.300E-01	1.68E-02	9.90E-02	7.83E-02	1.94E-01	1.77E-01
2.400E-01	1.55E-02	9.77E-02	6.92E-02	1.82E-01	1.67E-01
2.500E-01	1.44E-02	9.65E-02	6.16E-02	1.72E-01	1.58E-01
2.600E-01	1.33E-02	9.54E-02	5.50E-02	1.64E-01	1.50E-01
2.700E-01	1.24E-02	9.42E-02	4.94E-02	1.56E-01	1.44E-01
2.800E-01	1.16E-02	9.32E-02	4.45E-02	1.49E-01	1.38E-01
2.900E-01	1.08E-02	9.21E-02	4.03E-02	1.43E-01	1.32E-01
3.000E-01	1.01E-02	9.11E-02	3.67E-02	1.38E-01	1.28E-01
3.100E-01	9.52E-03	9.01E-02	3.34E-02	1.33E-01	1.24E-01
3.200E-01	8.96E-03	8.91E-02	3.06E-02	1.29E-01	1.20E-01
3.300E-01	8.44E-03	8.82E-02	2.81E-02	1.25E-01	1.16E-01
3.400E-01	7.97E-03	8.73E-02	2.59E-02	1.21E-01	1.13E-01
3.500E-01	7.53E-03	8.64E-02	2.39E-02	1.18E-01	1.10E-01
3.600E-01	7.14E-03	8.55E-02	2.21E-02	1.15E-01	1.08E-01

Molybdenum

Photon Energy (MeV)	coh (cm2/g)	Comp (cm2/g)	PE (cm2/g)	TOTAL w/coh (cm2/g)	TOTAL w/o coh (cm2/g)
3.700E-01	6.77E-03	8.47E-02	2.05E-02	1.12E-01	1.05E-01
3.800E-01	6.43E-03	8.39E-02	1.91E-02	1.09E-01	1.03E-01
3.900E-01	6.11E-03	8.31E-02	1.78E-02	1.07E-01	1.01E-01
4.000E-01	5.82E-03	8.23E-02	1.66E-02	1.05E-01	9.89E-02
4.100E-01	5.54E-03	8.15E-02	1.55E-02	1.03E-01	9.71E-02
4.200E-01	5.29E-03	8.08E-02	1.46E-02	1.01E-01	9.54E-02
4.300E-01	5.05E-03	8.01E-02	1.37E-02	9.88E-02	9.38E-02
4.400E-01	4.83E-03	7.94E-02	1.29E-02	9.71E-02	9.23E-02
4.500E-01	4.62E-03	7.87E-02	1.22E-02	9.54E-02	9.08E-02
4.600E-01	4.43E-03	7.80E-02	1.15E-02	9.39E-02	8.95E-02
4.700E-01	4.25E-03	7.73E-02	1.09E-02	9.24E-02	8.82E-02
4.800E-01	4.08E-03	7.67E-02	1.03E-02	9.11E-02	8.70E-02
4.900E-01	3.92E-03	7.61E-02	9.75E-03	8.97E-02	8.58E-02
5.000E-01	3.76E-03	7.54E-02	9.26E-03	8.85E-02	8.47E-02

Nickel
 Constituents (Atomic Number:Fraction by Weight)
  28:1.00000

 Partial Interaction Coefficients and Total Attenuation Coefficients

Photon Energy (MeV)	coh (cm2/g)	Comp (cm2/g)	PE (cm2/g)	TOTAL w/coh (cm2/g)	w/o coh (cm2/g)
2.000E-02	5.95E-01	1.16E-01	3.15E+01	3.22E+01	3.16E+01
3.000E-02	3.30E-01	1.30E-01	9.88E+00	1.03E+01	1.00E+01
4.000E-02	2.08E-01	1.36E-01	4.26E+00	4.60E+00	4.39E+00
5.000E-02	1.44E-01	1.38E-01	2.19E+00	2.47E+00	2.33E+00
6.000E-02	1.06E-01	1.38E-01	1.27E+00	1.51E+00	1.41E+00
7.000E-02	8.17E-02	1.37E-01	7.95E-01	1.01E+00	9.32E-01
8.000E-02	6.48E-02	1.36E-01	5.30E-01	7.31E-01	6.66E-01
9.000E-02	5.27E-02	1.34E-01	3.70E-01	5.57E-01	5.04E-01
1.000E-01	4.37E-02	1.32E-01	2.68E-01	4.44E-01	4.00E-01
1.100E-01	3.68E-02	1.30E-01	2.00E-01	3.67E-01	3.31E-01
1.200E-01	3.13E-02	1.28E-01	1.53E-01	3.13E-01	2.82E-01
1.300E-01	2.70E-02	1.26E-01	1.20E-01	2.73E-01	2.46E-01
1.400E-01	2.36E-02	1.25E-01	9.57E-02	2.44E-01	2.20E-01
1.500E-01	2.07E-02	1.23E-01	7.75E-02	2.21E-01	2.00E-01
1.600E-01	1.84E-02	1.21E-01	6.36E-02	2.03E-01	1.84E-01
1.700E-01	1.64E-02	1.19E-01	5.29E-02	1.88E-01	1.72E-01
1.800E-01	1.47E-02	1.17E-01	4.44E-02	1.76E-01	1.62E-01
1.900E-01	1.33E-02	1.16E-01	3.77E-02	1.67E-01	1.53E-01
2.000E-01	1.20E-02	1.14E-01	3.23E-02	1.58E-01	1.46E-01
2.100E-01	1.10E-02	1.12E-01	2.79E-02	1.51E-01	1.40E-01
2.200E-01	1.00E-02	1.11E-01	2.42E-02	1.45E-01	1.35E-01
2.300E-01	9.20E-03	1.09E-01	2.12E-02	1.40E-01	1.31E-01
2.400E-01	8.48E-03	1.08E-01	1.87E-02	1.35E-01	1.27E-01
2.500E-01	7.84E-03	1.06E-01	1.66E-02	1.31E-01	1.23E-01
2.600E-01	7.27E-03	1.05E-01	1.47E-02	1.27E-01	1.20E-01
2.700E-01	6.76E-03	1.04E-01	1.32E-02	1.24E-01	1.17E-01
2.800E-01	6.30E-03	1.03E-01	1.19E-02	1.21E-01	1.14E-01
2.900E-01	5.88E-03	1.01E-01	1.07E-02	1.18E-01	1.12E-01
3.000E-01	5.51E-03	1.00E-01	9.70E-03	1.15E-01	1.10E-01
3.100E-01	5.17E-03	9.90E-02	8.82E-03	1.13E-01	1.08E-01
3.200E-01	4.86E-03	9.79E-02	8.05E-03	1.11E-01	1.06E-01
3.300E-01	4.58E-03	9.69E-02	7.38E-03	1.09E-01	1.04E-01
3.400E-01	4.32E-03	9.58E-02	6.77E-03	1.07E-01	1.03E-01
3.500E-01	4.08E-03	9.48E-02	6.24E-03	1.05E-01	1.01E-01
3.600E-01	3.86E-03	9.38E-02	5.76E-03	1.03E-01	9.96E-02

Nickel

Photon Energy (MeV)	coh (cm2/g)	Comp (cm2/g)	PE (cm2/g)	TOTAL w/coh (cm2/g)	w/o coh (cm2/g)
3.700E-01	3.66E-03	9.29E-02	5.34E-03	1.02E-01	9.82E-02
3.800E-01	3.48E-03	9.20E-02	4.96E-03	1.00E-01	9.69E-02
3.900E-01	3.30E-03	9.11E-02	4.61E-03	9.90E-02	9.57E-02
4.000E-01	3.14E-03	9.02E-02	4.30E-03	9.76E-02	9.45E-02
4.100E-01	2.99E-03	8.94E-02	4.02E-03	9.64E-02	9.34E-02
4.200E-01	2.86E-03	8.85E-02	3.76E-03	9.51E-02	9.23E-02
4.300E-01	2.73E-03	8.77E-02	3.53E-03	9.40E-02	9.12E-02
4.400E-01	2.61E-03	8.69E-02	3.32E-03	9.29E-02	9.02E-02
4.500E-01	2.49E-03	8.62E-02	3.12E-03	9.18E-02	8.93E-02
4.600E-01	2.39E-03	8.54E-02	2.94E-03	9.07E-02	8.84E-02
4.700E-01	2.29E-03	8.47E-02	2.78E-03	8.98E-02	8.75E-02
4.800E-01	2.20E-03	8.40E-02	2.63E-03	8.88E-02	8.66E-02
4.900E-01	2.11E-03	8.33E-02	2.49E-03	8.79E-02	8.58E-02
5.000E-01	2.03E-03	8.26E-02	2.37E-03	8.70E-02	8.50E-02

N5 (nickel superalloy)
    Constituents (Atomic Number:Fraction by Weight)
        13:0.06200 24:0.07000 27:0.07500 28:0.63300 42:0.01500 73:0.06500
        74:0.05000 75:0.03000

Partial Interaction Coefficients and Total Attenuation Coefficients

Photon Energy (MeV)	coh (cm2/g)	Comp (cm2/g)	PE (cm2/g)	TOTAL w/coh (cm2/g)	w/o coh (cm2/g)
2.000E-02	7.70E-01	1.11E-01	3.39E+01	3.48E+01	3.40E+01
3.000E-02	4.38E-01	1.24E-01	1.09E+01	1.14E+01	1.10E+01
4.000E-02	2.81E-01	1.30E-01	4.76E+00	5.17E+00	4.89E+00
5.000E-02	1.96E-01	1.32E-01	2.49E+00	2.82E+00	2.62E+00
6.000E-02	1.46E-01	1.32E-01	1.46E+00	1.74E+00	1.59E+00
7.000E-02	1.13E-01	1.32E-01	1.89E+00	2.14E+00	2.02E+00
8.000E-02	9.00E-02	1.31E-01	1.49E+00	1.71E+00	1.62E+00
9.000E-02	7.35E-02	1.29E-01	1.08E+00	1.28E+00	1.21E+00
1.000E-01	6.11E-02	1.28E-01	8.10E-01	9.99E-01	9.38E-01
1.100E-01	5.16E-02	1.26E-01	6.23E-01	8.00E-01	7.49E-01
1.200E-01	4.41E-02	1.24E-01	4.89E-01	6.57E-01	6.13E-01
1.300E-01	3.81E-02	1.22E-01	3.91E-01	5.51E-01	5.13E-01
1.400E-01	3.33E-02	1.20E-01	3.18E-01	4.72E-01	4.38E-01
1.500E-01	2.94E-02	1.18E-01	2.62E-01	4.10E-01	3.81E-01
1.600E-01	2.61E-02	1.17E-01	2.19E-01	3.62E-01	3.36E-01
1.700E-01	2.33E-02	1.15E-01	1.85E-01	3.23E-01	3.00E-01
1.800E-01	2.10E-02	1.13E-01	1.58E-01	2.92E-01	2.71E-01
1.900E-01	1.90E-02	1.12E-01	1.36E-01	2.66E-01	2.47E-01
2.000E-01	1.72E-02	1.10E-01	1.18E-01	2.45E-01	2.28E-01
2.100E-01	1.57E-02	1.09E-01	1.03E-01	2.27E-01	2.11E-01
2.200E-01	1.44E-02	1.07E-01	9.04E-02	2.12E-01	1.98E-01
2.300E-01	1.33E-02	1.06E-01	8.00E-02	1.99E-01	1.86E-01
2.400E-01	1.22E-02	1.04E-01	7.12E-02	1.88E-01	1.76E-01
2.500E-01	1.13E-02	1.03E-01	6.36E-02	1.78E-01	1.67E-01
2.600E-01	1.05E-02	1.02E-01	5.72E-02	1.70E-01	1.59E-01
2.700E-01	9.80E-03	1.01E-01	5.16E-02	1.62E-01	1.52E-01
2.800E-01	9.15E-03	9.93E-02	4.68E-02	1.55E-01	1.46E-01
2.900E-01	8.56E-03	9.82E-02	4.26E-02	1.49E-01	1.41E-01
3.000E-01	8.02E-03	9.71E-02	3.88E-02	1.44E-01	1.36E-01
3.100E-01	7.54E-03	9.60E-02	3.56E-02	1.39E-01	1.32E-01
3.200E-01	7.09E-03	9.49E-02	3.27E-02	1.35E-01	1.28E-01
3.300E-01	6.69E-03	9.39E-02	3.01E-02	1.31E-01	1.24E-01
3.400E-01	6.31E-03	9.29E-02	2.79E-02	1.27E-01	1.21E-01
3.500E-01	5.97E-03	9.19E-02	2.58E-02	1.24E-01	1.18E-01

N5 (nickel superalloy)

Photon Energy (MeV)	coh (cm2/g)	Comp (cm2/g)	PE (cm2/g)	TOTAL w/coh (cm2/g)	w/o coh (cm2/g)
3.600E-01	5.66E-03	9.10E-02	2.40E-02	1.21E-01	1.15E-01
3.700E-01	5.37E-03	9.01E-02	2.23E-02	1.18E-01	1.12E-01
3.800E-01	5.10E-03	8.92E-02	2.08E-02	1.15E-01	1.10E-01
3.900E-01	4.85E-03	8.83E-02	1.95E-02	1.13E-01	1.08E-01
4.000E-01	4.62E-03	8.75E-02	1.83E-02	1.10E-01	1.06E-01
4.100E-01	4.40E-03	8.66E-02	1.71E-02	1.08E-01	1.04E-01
4.200E-01	4.20E-03	8.58E-02	1.61E-02	1.06E-01	1.02E-01
4.300E-01	4.01E-03	8.51E-02	1.52E-02	1.04E-01	1.00E-01
4.400E-01	3.84E-03	8.43E-02	1.43E-02	1.02E-01	9.86E-02
4.500E-01	3.67E-03	8.36E-02	1.36E-02	1.01E-01	9.71E-02
4.600E-01	3.52E-03	8.28E-02	1.28E-02	9.92E-02	9.57E-02
4.700E-01	3.38E-03	8.21E-02	1.22E-02	9.77E-02	9.43E-02
4.800E-01	3.24E-03	8.15E-02	1.15E-02	9.62E-02	9.30E-02
4.900E-01	3.11E-03	8.08E-02	1.10E-02	9.49E-02	9.18E-02
5.000E-01	2.99E-03	8.01E-02	1.04E-02	9.36E-02	9.06E-02

Tungsten
Constituents (Atomic Number:Fraction by Weight)
74:1.00000

Partial Interaction Coefficients and Total Attenuation Coefficients

Photon Energy (MeV)	coh (cm2/g)	Comp (cm2/g)	PE (cm2/g)	TOTAL w/coh (cm2/g)	w/o coh (cm2/g)
2.000E-02	2.04E+00	7.25E-02	6.36E+01	6.57E+01	6.37E+01
3.000E-02	1.20E+00	8.64E-02	2.14E+01	2.27E+01	2.15E+01
4.000E-02	7.94E-01	9.43E-02	9.78E+00	1.07E+01	9.87E+00
5.000E-02	5.61E-01	9.86E-02	5.29E+00	5.95E+00	5.39E+00
6.000E-02	4.21E-01	1.01E-01	3.19E+00	3.71E+00	3.29E+00
7.000E-02	3.28E-01	1.02E-01	1.06E+01	1.10E+01	1.07E+01
8.000E-02	2.64E-01	1.03E-01	7.44E+00	7.81E+00	7.54E+00
9.000E-02	2.17E-01	1.02E-01	5.47E+00	5.79E+00	5.57E+00
1.000E-01	1.81E-01	1.02E-01	4.15E+00	4.44E+00	4.26E+00
1.100E-01	1.54E-01	1.01E-01	3.23E+00	3.48E+00	3.33E+00
1.200E-01	1.32E-01	1.00E-01	2.55E+00	2.79E+00	2.66E+00
1.300E-01	1.15E-01	9.95E-02	2.06E+00	2.27E+00	2.16E+00
1.400E-01	1.00E-01	9.85E-02	1.68E+00	1.88E+00	1.78E+00
1.500E-01	8.88E-02	9.74E-02	1.40E+00	1.58E+00	1.49E+00
1.600E-01	7.91E-02	9.63E-02	1.17E+00	1.35E+00	1.27E+00
1.700E-01	7.09E-02	9.52E-02	9.93E-01	1.16E+00	1.09E+00
1.800E-01	6.40E-02	9.40E-02	8.51E-01	1.01E+00	9.45E-01
1.900E-01	5.80E-02	9.29E-02	7.35E-01	8.86E-01	8.28E-01
2.000E-01	5.29E-02	9.18E-02	6.40E-01	7.84E-01	7.32E-01
2.100E-01	4.84E-02	9.08E-02	5.61E-01	7.00E-01	6.52E-01
2.200E-01	4.45E-02	8.97E-02	4.95E-01	6.29E-01	5.84E-01
2.300E-01	4.10E-02	8.87E-02	4.39E-01	5.69E-01	5.28E-01
2.400E-01	3.79E-02	8.77E-02	3.92E-01	5.17E-01	4.79E-01
2.500E-01	3.52E-02	8.67E-02	3.51E-01	4.73E-01	4.38E-01
2.600E-01	3.27E-02	8.57E-02	3.16E-01	4.35E-01	4.02E-01
2.700E-01	3.05E-02	8.48E-02	2.86E-01	4.01E-01	3.71E-01
2.800E-01	2.86E-02	8.39E-02	2.60E-01	3.72E-01	3.44E-01
2.900E-01	2.68E-02	8.30E-02	2.37E-01	3.47E-01	3.20E-01
3.000E-01	2.51E-02	8.21E-02	2.17E-01	3.24E-01	2.99E-01
3.100E-01	2.36E-02	8.13E-02	1.99E-01	3.04E-01	2.80E-01
3.200E-01	2.23E-02	8.05E-02	1.83E-01	2.86E-01	2.63E-01
3.300E-01	2.10E-02	7.97E-02	1.69E-01	2.70E-01	2.49E-01
3.400E-01	1.99E-02	7.89E-02	1.56E-01	2.55E-01	2.35E-01
3.500E-01	1.88E-02	7.81E-02	1.45E-01	2.42E-01	2.23E-01
3.600E-01	1.78E-02	7.74E-02	1.35E-01	2.30E-01	2.12E-01

Tungsten

Photon Energy (MeV)	coh (cm2/g)	Comp (cm2/g)	PE (cm2/g)	TOTAL w/coh (cm2/g)	w/o coh (cm2/g)
3.700E-01	1.69E-02	7.66E-02	1.26E-01	2.19E-01	2.02E-01
3.800E-01	1.61E-02	7.59E-02	1.18E-01	2.10E-01	1.94E-01
3.900E-01	1.53E-02	7.53E-02	1.10E-01	2.01E-01	1.85E-01
4.000E-01	1.46E-02	7.46E-02	1.03E-01	1.92E-01	1.78E-01
4.100E-01	1.39E-02	7.39E-02	9.71E-02	1.85E-01	1.71E-01
4.200E-01	1.33E-02	7.33E-02	9.14E-02	1.78E-01	1.65E-01
4.300E-01	1.27E-02	7.27E-02	8.62E-02	1.72E-01	1.59E-01
4.400E-01	1.22E-02	7.21E-02	8.14E-02	1.66E-01	1.53E-01
4.500E-01	1.17E-02	7.15E-02	7.71E-02	1.60E-01	1.49E-01
4.600E-01	1.12E-02	7.09E-02	7.30E-02	1.55E-01	1.44E-01
4.700E-01	1.07E-02	7.03E-02	6.93E-02	1.50E-01	1.40E-01
4.800E-01	1.03E-02	6.97E-02	6.58E-02	1.46E-01	1.36E-01
4.900E-01	9.91E-03	6.92E-02	6.26E-02	1.42E-01	1.32E-01
5.000E-01	9.54E-03	6.87E-02	5.96E-02	1.38E-01	1.28E-01

10.2 Physical Properties of Metals Used in the XRD Experiments

Metal	Atomic Number	Density (kg m⁻³) at 20 °C	Density at Solidus (kg m⁻³)	Density at Liquidus (kg m⁻³)	Melting Temperature (°C)
Aluminum (Al)	13	2.7×10^3	2.567×10^3	2.385×10^3	660
Copper (Cu)	29	8.96×10^3	8.404×10^3	8.000×10^3	1083
Gallium (Ga)	31	5.91×10^3	5.91×10^3	6.09×10^3	29.8
Nickel (Ni)	28	8.9×10^3	8.304×10^3	7.905×10^3	1455

10.3 Composition of N5 Nickel Alloy

In weight percent, this is the approximate composition for the major elements.

7.0%   Cr
7.5%   Co
1.5%   Mo
6.2%   Al
6.5%   Ta
5.0%   W
3.0%   Re
Balance is Ni

10.4 Anomalous Dispersion Corrections for Nickel

The factors f1 and f2, for nickel, as a function of energy are tabulated below. The source of this data is reference [25].

Ni: Z = 28
Atomic weight =58.69000 g/mol
Nominal density:8.876g/cm3

μa(barns/atom) = μ(cm2/g) x 97.4571
E(eV) μ(cm2/g) = f2(e/atom) x716993

Location (keV) of Absorption Edges:
K   8.33280E+00 L I   1.00810E+00 L II 8.71900E-01 L III 8.54700E-01
M I   1.11800E-01 M II 6.81000E-02 M III 6.81000E-02 M IV 3.60000E-03
M V   3.60000E-03

Relativistic correction estimate frel (H82,3/5CL) =   -0.12883   -0.081 e/atom
Nuclear Thomson correction fNT =   -0.0073281   e/atom

E (keV)	f1 (e/atom)	f2 (e/atom)	E (keV)	f1 (e/atom)	f2 (e/atom)
2.080733	27.9855	5.1095	7.902609	25.4382	0.51866
2.224304	28.0289	4.592	8.166144	24.5187	0.49051
2.377781	28.0405	4.1285	8.291136	23.0862	0.47804
2.541848	28.0299	3.7138	8.324467	21.3577	0.4748
2.717235	28.0016	3.343	8.374464	23.1229	3.8783
2.904724	27.9659	3.0114	8.44789	24.2866	3.8127
3.10515	27.9386	2.6981	8.499456	24.7181	3.7677
3.319406	27.8847	2.4055	9.030794	26.3786	3.4017
3.548445	27.8165	2.1392	9.653919	27.1216	3.0709
3.793288	27.734	1.897	10.32004	27.5654	2.7492
4.055024	27.6425	1.6844	11.03212	27.8453	2.4485
4.334821	27.5458	1.4976	11.79334	28.0311	2.1838
4.633924	27.4435	1.3269	12.60708	28.1583	1.9478
4.953664	27.3322	1.1761	13.47697	28.2454	1.7376
5.295467	27.2114	1.0433	14.40688	28.3043	1.5502
5.660855	27.0779	0.92645	15.40095	28.3432	1.3831
6.051453	26.9254	0.82342	16.46362	28.3682	1.2341
6.469004	26.7418	0.73254	17.59961	28.4192	1.0942
6.915365	26.5023	0.65229	18.81398	28.4194	0.96902
7.392525	26.1454	0.58138	20.11215	28.412	0.8582

E (keV)	f1 (e/atom)	f2 (e/atom)	E (keV)	f1 (e/atom)	f2 (e/atom)
21.500	28.400	0.76013	121.859	28.043	0.024598
22.983	28.384	0.67282	130.267	28.039	0.021438
24.569	28.366	0.59559	139.255	28.035	0.018685
26.265	28.348	0.52725	148.864	28.031	0.016286
28.077	28.329	0.46461	159.136	28.028	0.014196
30.014	28.309	0.40881	170.116	28.025	0.012374
32.085	28.288	0.35972	181.854	28.022	0.010786
34.299	28.268	0.31655	194.402	28.020	0.009402
36.666	28.248	0.27821	207.816	28.018	0.008196
39.195	28.230	0.24347	222.155	28.016	0.0071448
41.900	28.211	0.21307	237.484	28.014	0.0062286
44.791	28.193	0.18647	253.870	28.013	0.00543
47.882	28.177	0.1632	271.387	28.011	0.0047339
51.185	28.161	0.14284	290.113	28.010	0.0041272
54.717	28.147	0.12502	310.130	28.009	0.0035983
58.493	28.133	0.10943	331.529	28.008	0.0031373
62.529	28.121	0.095778	354.405	28.007	0.0027354
66.843	28.110	0.083836	378.859	28.006	0.002385
71.455	28.099	0.073385	405.000	28.006	0.0020796
76.386	28.090	0.064239	432.945	28.005	0.0018133
81.656	28.081	0.056155			
87.291	28.073	0.048934			
93.314	28.066	0.042642			
99.752	28.060	0.037161			
106.635	28.054	0.032385			
113.993	28.048	0.028224			

10.5 Model of Transmission XRD from a Sample Surrounded by Mold Material

Example: Nickel sample with Refractory Oxide Mold Walls   File: XRDEFNI3.MCD

$IO := 10^6$ $W1 := 0.6$ $W2 := 0.6$ $NL := 1000$ $L := 1.0$ $dx := \dfrac{L}{NL}$

$\rho s := 8.9$ density of Ni (g/cm³)

$\rho w := 2.46$ density of Mold Material (g/cm³)

K is the forward scattering factor, the fraction of coherently scattered photons scattered into the area occupied by the x-ray imager.

D is the imager diameter, SID is the source-to-imager distance

$D := 50$ $SID := 150$

$\theta := atan\left(\dfrac{\dfrac{D}{2}}{SID}\right)$ $\theta = 0.165$

$R := \dfrac{SID}{\cos(\theta)}$ $R = 152.069$ $h := R - SID$ $h = 2.069$

$K := \dfrac{h}{2 \cdot R}$ $K = 6.803 \cdot 10^{-3}$

$i := 1 .. 49$

Energy (keV) $E_i := READ("energys.dat")$ Coherent Atten in Sample $\mu sr_i := READ("nicoh.dat")$

Total Atten in Sample $\mu st_i := READ("nitot.dat")$ Total Atten in Mold Wall $\mu wt_i := READ("hmoldtot.dat")$

Attenuation coefficients are mass attenuation coefficients (cm²/g). Multiply these by the density to find the linear attenuation coefficients.

MODEL OF RADIATION TRANSPORT
IN X-RAY DIFFRACTION SENSOR EXPERIMENTS

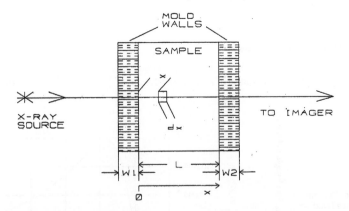

I0$_i$ is the intensity (at energy E$_i$) of the x-ray beam, with no sample or mold present.

I1$_i$ is the intensity transmitted through the entrance mold wall.

$$I1_i := I0 \cdot e^{-\mu wt_i \cdot \rho w \cdot W1} \qquad\qquad\qquad\qquad (A1)$$

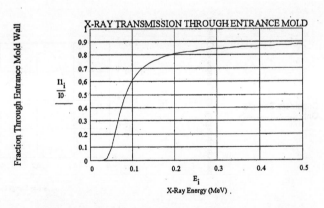

I2$_i$ is the intensity of coherent (Rayleigh) scatter which exits from the sample. It is a integral of the product of (1) the attenuation in the sample up to the point of coherent scatter, (2) the efficiency of generation of coherent scatter, and (3) the attenuation of the coherent scatter as it emerges from the sample.

123

$$I2_i := I1_i \cdot \sum_{n=1}^{NL} e^{-\mu st_i \cdot \rho s \cdot n \cdot dx} \cdot K \cdot \left(1 - e^{-\mu sr_i \cdot \rho s \cdot dx}\right) \cdot e^{-\mu st_i \cdot \rho s \cdot (L - (n \cdot dx))} \qquad \text{(A2)}$$

Symbolically simplfying eq. (2) yields:

$$I2_i := -I1_i \cdot K \cdot NL \cdot \left[-\exp\left(-\mu st_i \cdot \rho s \cdot L\right) + \exp\left[-\rho s \cdot \left(\mu st_i \cdot L + \mu sr_i \cdot dx\right)\right] \right] \qquad \text{(A3)}$$

Considers Sample
& Entrance Mold -->

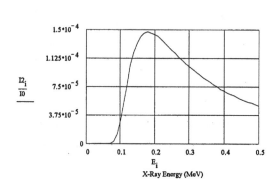

Considers Sample
Only ----,------->

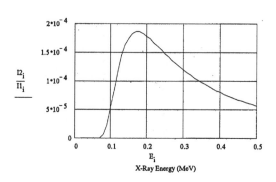

I3(E) is the intensity of coherent scatter which is transmitted through the exit mold wall.

$$I3_i := I2_i \cdot e^{-\mu wt_i \cdot \rho w \cdot W2} \qquad \text{(A4)}$$

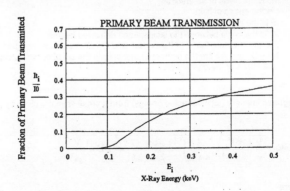

WRITE("ni6100h6.dat") $:= \dfrac{I3_i}{I0}$

Calculate the intensity in the primary beam transmitted through both mold walls and the sample. The transmitted primary beam is what would normally be used to form a radiographic image.

$$It_i := I0 \cdot e^{-\mu wt_i \cdot \rho w \cdot W1} \cdot e^{-\mu st_i \cdot \rho s \cdot L} \cdot e^{-\mu wt_i \cdot \rho w \cdot W2} \qquad \text{(A5)}$$

The plots are for a 6 mm entrance mold, 10 mm Ni sample, and a 6 mm exit mold.

125

Fundamental Physics Model

As an x-ray beam travels through the mold and a crystalline specimen, attenuation and diffraction will occur.

Attenuation of a narrow, monochromatic x-ray beam follows the relation

$$I := I_0 \cdot e^{\frac{-\mu}{\rho} \cdot \rho \cdot x} \qquad \blacksquare$$

(1)

where
    I is the intensity after passing through the material,
    I_0 is the incident intensity,
    μ/ρ is the mass attenuation coefficient (material and energy dependent),
    ρ is the density of the material, and
    x is the thickness of the material.

The intensity of x-rays diffracted by a small crystal is given by [Warren, eqn (3.6), p 29]

$$I_d := I_0 \cdot \left(\frac{e0^4}{R^2 \cdot m0^2 \cdot c^4}\right) \cdot \left(\frac{1}{4 \cdot \pi \cdot \varepsilon 0}\right)^2 \cdot \frac{\left[1 + \cos^2 (2 \cdot \theta)\right]}{2} \cdot F \cdot F' \cdot N1^2 \cdot N2^2 \cdot N3^2 \qquad \blacksquare$$

(2)

where
    I_d is the diffracted intensity,
    I_0 is the incident intensity,
    $e0$ is the charge of an electron,
    R is the distance from the origin of diffraction to a point of observation,
    $m0$ is the mass of an electron,
    c is the speed of light,
    $\varepsilon 0$ is the permittivity of free space,
    2θ is the angle between the incident and diffracted beams,
    F is the structure factor of the crystal, and F' is its complex conjugate,
    N1, N2, and N3 are the number of unit cells in the small crystal (along the
    directions of the crystal axes).

The integrated intensity (total diffracted energy) from a small crystal is given by [Warren, eqn (4.6), p 44]

$$DiffEn := \frac{I_0}{\omega} \cdot \left(\frac{e0^4}{m0^2 \cdot c^4}\right) \cdot \left(\frac{1}{4 \cdot \pi \cdot \varepsilon 0}\right)^2 \cdot \frac{\left[1 + \cos^2 (2 \cdot \theta)\right]}{2 \cdot \sin(2 \cdot \theta)} \cdot F \cdot F' \cdot \lambda^3 \cdot \frac{V}{va^2} \qquad \blacksquare$$

(3)

where
    ω is the angular velocity that the primary beam rotates (required to integrate through
    the area under the interference function sin2(Ny)/sin2(y),
    V is the volume of the small crystal, and
    va is the volume of the unit cell.

126

Physical Constants

Planck's constant

$h := 6.6260755 \cdot 10^{-34} \cdot \text{joule} \cdot \text{sec}$

Mass of an Electron

$m0 := 9.1093897 \cdot 10^{-31} \cdot \text{kg}$

Speed of Light

$c := 299792458 \cdot \dfrac{m}{\text{sec}}$

Permittivity of vacuum

$\epsilon0 := 8.854187817 \cdot 10^{-12} \cdot \dfrac{\text{farad}}{m}$

Charge of an Electron

$e0 := 1.60217733 \cdot 10^{-19} \cdot \text{coul}$

$\text{Energy} := m0 \cdot c^2$

$\text{Energy} = 8.18711 \cdot 10^{-14} \cdot \text{kg} \cdot m^2 \cdot \text{sec}^{-2}$

Example: Single-Crystal Nickel Specimen Surrounded by Refractory Oxide Mold Walls

Mold wall thicknesses (m) $W1 := 0.$ $W2 := 0.$

Density of Mold Material (kg/m³) $\rho w := 2.46 \cdot 10^3$

Thickness of specimen (m) $L := 10. \cdot 10^{-3}$

Density of Ni (kg/m³) $\rho s := 8.9 \cdot 10^3$

Divide the specimen into a number of segments $NL := 10000$

Length of a segment (cm) $\Delta x := \dfrac{L}{NL}$ $\Delta x = 1 \cdot 10^{-6}$

Lattice constant (length of an edge in the unit cell) for Ni (m) $a := 3.524 \cdot 10^{-10}$

Atomic Number of Nickel $Z := 28$

Read in the energy points $i := 1 .. 49$

$E_i := \text{READ}(\text{"energys.dat"})$ Energy (MeV)

$E_i := E_i \cdot 1000$ Convert to (keV) $E_9 = 100$

$\lambda_i := \dfrac{12.398424468}{E_i}$ λ is in (Angstroms), for E in (keV) $\lambda_9 = 0.12398$

$\lambda_i := \lambda_i \cdot 10^{-10}$ Convert λ to (m). $\lambda_9 = 1.23984 \cdot 10^{-11}$

Read in the incident x-ray **source spectrum** (flux per energy interval) at each of the energy points (E_i).

$I0_i := \text{READ}(\text{"spect001.dat"})$

127

Read the **attenuation coefficients** at each energy point, for the mold material and the specimen.

Total Atten $\mu wt_i := \text{READ}(\text{"hmoldtot.dat"})$ Total Atten $\mu st_i := \text{READ}(\text{"nitot.dat"})$
in Mold Wall   in Sample

Attenuation coefficients are **mass attenuation coefficients** (cm²/g). Multiply
these by the density to get the linear attenuation coefficients.

$$\mu wt_i := \frac{\mu wt_i}{10} \qquad \mu st_i := \frac{\mu st_i}{10} \qquad \text{Convert the units to (m}^2\text{/kg).}$$

Choose a set of lattice planes in the crystal [Miller indices (hkl)]. This defines a particular "reflection"
to investigate. The indices should be unmixed (nickel is fcc) for F (structure factor) to be nonzero.

Let

$$h := 1 \qquad k := 1 \qquad l := 1$$

The spacing of lattice planes, d, for a crystal of cubic symmetry is

$$\frac{1}{d^2} := \frac{\left(h^2 + k^2 + l^2\right)}{a^2} \qquad \text{where} \qquad \text{a is the lattice constant.} \qquad (4)$$

Diffraction will occur when Bragg's
Law,

$$n \cdot \lambda := 2 \cdot d \cdot \sin(\theta) \qquad \text{is satisfied.} \qquad (5)$$

Substituting the expression for d (from (4)) into Bragg's Law gives the scattering angle θ for a cubic crystal:

$$\sin(\theta)^2 := \frac{\lambda^2}{4 \cdot a^2} \cdot \left(h^2 + k^2 + l^2\right)$$

Note that:

$$\sin\theta_i := \sqrt{\frac{(\lambda_i)^2}{4 \cdot a^2} \cdot \left(h^2 + k^2 + l^2\right)} \qquad \text{or} \qquad \sin\theta_i := \frac{1}{2} \cdot \frac{\lambda_i}{a} \cdot \sqrt{h^2 + k^2 + l^2} \qquad \frac{\sin\theta_i}{\lambda_i} := \frac{1}{2 \cdot a} \cdot \sqrt{h^2 + k^2 + l^2}$$

is independent of energy

1/2 Scattering $\theta_i := \text{asin}\left(\frac{1}{2} \cdot \frac{\lambda_i}{a} \cdot \sqrt{h^2 + k^2 + l^2}\right)$ (6)   Convert to degrees $\text{Theta}_i := \theta_i \cdot \frac{180}{\pi}$
Angle (θ)

$$\text{Theta}_9 = 1.74603$$

Scattering $\theta 2_i := 2 \cdot \theta_i$ radians
Angle, 2θ

128

2θ is the **scattering angle** for a given set of lattice planes and a given energy (wavelength) x-ray.

Diffraction will occur **only if** the correct energy x-ray is incident on the crystal at the correct angle.

For each energy considered, we assume the specimen is correctly aligned for diffraction.

$(2 \cdot \text{Theta})_i$
————

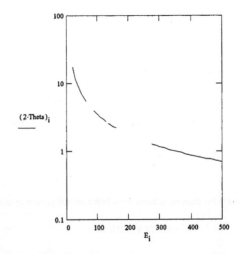

The structure factor **F**, is a function of the diffracting crystal material, its crystal structure, the particular diffracting planes in the crystal, the incident x-ray energy, and the diffracted angle.

$$F := \sum_{n=1}^{N} f_n \cdot \exp\left[i \cdot 2 \cdot \pi \cdot \left(h \cdot x_n + k \cdot y_n + l \cdot z_n\right)\right] \qquad (7)$$

where

   f_n is the atomic scattering factor of the nth atom in the unit cell
      (it is a function of angle and energy),
   (hkl) are the Miller indices of the diffracting planes,
   x,y,z are the relative positions of the (N) atoms in the unit cell.

For a face-centered cubic (fcc) crystal, F is zero if the indices (hkl) are mixed (even, odd) and is 4*f if the indices are unmixed.

The **atomic scattering factor, f,** is a complex number.

$$fReal := f0 + f1 + fNT + frel - Z \,\blacksquare \qquad\qquad fImag := f2 \,\blacksquare \qquad (8)$$

   f0 is a function of $\sin(\theta)/\lambda$, fNT and frel can be considered constants, and
   f1 and f2 are functions of energy
      [CT Chantler, Theoretical form factor, attenuation, and scaattering tabulation for
      Z=1-92 from E=1-10eV to E=0.4-1.0MeV, J. Chem. Ref. Data, Vol. 24, 71- (1995)]

129

Read in the coefficients for calculating the atomic scattering factor (f_0) for the specimen material.

An empirical, sum of exponentials, fit is used [D.. Waasmaier, A Kirfel, New analytical scattering factor functions for free atoms and ions, Acta. Cryst. A, 151, No. 3, 416-430 (1995)]. A_i, B_i, and C are the coefficients in the fit.

$$j := 1.. 11 \qquad e_j := READ("sfitni.dat")$$

$$j := 1.. 5 \qquad A_j := e_j \qquad A^T = [\, 0 \quad 13.52187 \quad 6.94728 \quad 3.86603 \quad 2.1359 \quad 4.28473 \,]$$

$$j := 6.. 10 \qquad B_{j-5} := e_j \qquad B^T = [\, 0 \quad 4.07728 \quad 0.28676 \quad 14.62263 \quad 71.96608 \quad 4.437 \cdot 10^{-3} \,]$$

$$C := e_{11} \qquad C = 2.7627$$

Read in the dispersion corrections to the atomic scattering factor.

$$f1_i := READ("f1vsENi.dat") \qquad\qquad f2_i := READ("f2vsENi.dat")$$

for Nickel ———> $\qquad fNT := -0.00733 \qquad frel := -8.1 \cdot 10^{-2}$

To determine the atomic scattering factor for the specimen, first calculate the argument of the function.

$$arg_i := \frac{\sin\theta_i}{\lambda_i \cdot 10^{10}} \qquad\qquad arg_9 = 0.24575$$

Note, λ was converted to units of Angstrom in the equation above

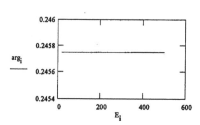

130

Calculate f0 using an empirical fit

$$f0_i := \sum_{j=1}^{5} A_j \cdot \exp\left[-B_j \cdot \left(arg_i\right)^2\right] + C \qquad\qquad f0_9 = 26.07104 \qquad\qquad (9)$$

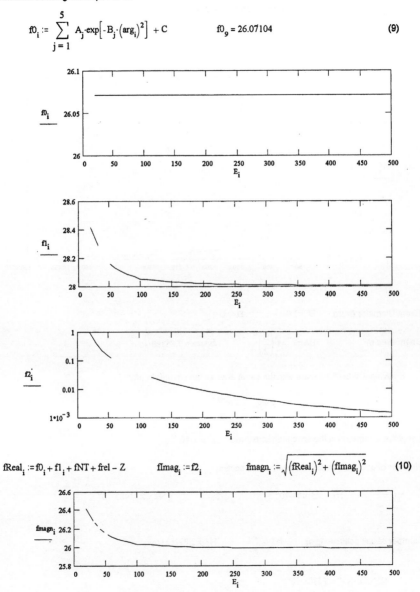

$$fReal_i := f0_i + f1_i + fNT + frel - Z \qquad flmag_i := f2_i \qquad fmagn_i := \sqrt{\left(fReal_i\right)^2 + \left(flmag_i\right)^2} \qquad (10)$$

131

The structure factor for an fcc crystal, and lattice planes defined by unmixed indices (hkl) is 4f . For intensities, square the amplitude structure factor, (that is, 16 f²).

$$FF'_i := 16 \cdot \left(fmagn_i\right)^2 \qquad (11)$$

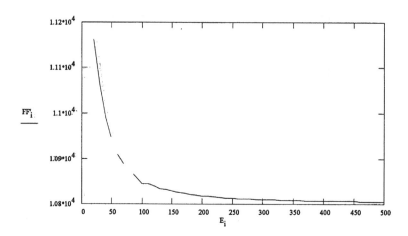

Beam Diameter (mm) $\quad D' := 1.0 \qquad D := D \cdot 10^{-3}$

Beam area m² $\qquad BArea := \pi \cdot \left(\dfrac{D}{2}\right)^2 \qquad BArea = 7.85398 \cdot 10^{-7}$

Equivalent side of a square with the same area as the circular beam

$$g := \sqrt{BArea} \qquad g = 8.86227 \cdot 10^{-4} \quad m$$

Length of a segment in the specimen is dx (m). $\quad \Delta x = 1 \cdot 10^{-6}$

Number of unit cells across the beam cross section

$$N1 := \frac{g}{a} \qquad N1 = 2.51483 \cdot 10^6 \qquad N2 := \frac{g}{a} \qquad N2 = 2.51483 \cdot 10^6$$

Number of unit cells along dx $\qquad N3 := \dfrac{\Delta x}{a} \qquad N3 = 2.83768 \cdot 10^3$

$$N1^2 \cdot N2^2 \cdot N3^2 = 3.2208 \cdot 10^{32}$$

132

Calculate the other quantities in eqn (3) to find the diffracted intensities.

$$\left(\frac{e0^4}{m0^2 \cdot c^4}\right) \cdot \left(\frac{1}{4 \cdot \pi \cdot \varepsilon 0}\right)^2 = 7.94079 \cdot 10^{-30} \cdot m^2$$

$$LP_i := \frac{\left[1 + \left(\cos\left(2 \cdot \theta_i\right)\right)^2\right]}{2 \cdot \sin\left(2 \cdot \theta_i\right)}$$

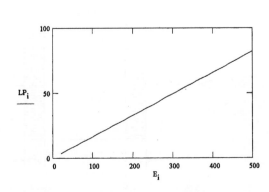

Let $\omega := 1$

$$V := BArea \cdot \Delta x \qquad V = 7.85398 \cdot 10^{-13} \qquad va := a^3 \qquad va = 4.37631 \cdot 10^{-29}$$

$$DiffEn_i := \frac{I0_i}{\omega} \cdot \left(\frac{e0^4}{m0^2 \cdot c^4}\right) \cdot \left(\frac{1}{4 \cdot \pi \cdot \varepsilon 0}\right)^2 \cdot \frac{\left(1 + \cos\left(2 \cdot \theta_i\right)^2\right)}{2 \cdot \sin\left(2 \cdot \theta_i\right)} \cdot FF_i \cdot \left(\lambda_i\right)^3 \cdot \frac{V}{va^2} \tag{12}$$

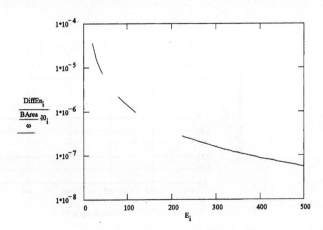

MODEL OF TRANSMISSION X-RAY DIFFRACTION
FOR A MOLD-ENCASED SINGLE-CRYSTAL
SPECIMEN

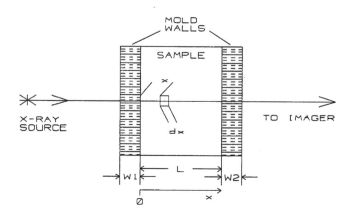

I0$_i$ is the intensity (at energy E$_i$) of the x-ray beam, with no sample or mold present.

I1$_i$ is the intensity transmitted through the entrance mold wall.

$$I1_i := I0_i \cdot \exp\left(-\mu wt_i \cdot \rho w \cdot W1\right)$$

(13)

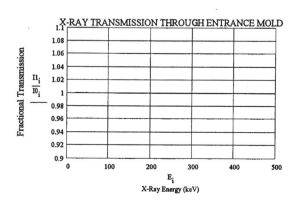

134

E2$_i$ is the total diffracted energy which emerges from the sample. It is the product of (1) the attenuation in the sample up to the point of diffraction, (2) the efficiency of diffraction generation, and (3) the attenuation of the diffracted x-rays as they exit through the remaining portion of the specimen. These contributions from each segment of the specimen are summed over the path of the incident x-ray beam through the specimen.

$$E2_i := I1_i \cdot \sum_{n=0}^{NL} \frac{e^{(-\mu st_i \cdot \rho s) \cdot n \cdot \Delta x}}{\omega} \cdot \left(\frac{e0^4}{m0^2 \cdot c^4}\right) \cdot \left(\frac{1}{4 \cdot \pi \cdot \varepsilon 0}\right)^2 \cdot \frac{\left(1 + \cos\left(2 \cdot \theta_i\right)^2\right)}{2 \cdot \sin\left(2 \cdot \theta_i\right)} \cdot FF'_i \cdot \left(\lambda_i\right)^3 \cdot \frac{BArea \cdot \Delta x}{va^2} \cdot e^{-\mu st_i \cdot \rho s \cdot (L - (n \cdot \Delta x))}$$

or

$$E2_i := I1_i \cdot \int_0^L \frac{e^{-\mu st_i \cdot \rho s \cdot x}}{\omega} \cdot \left(\frac{e0^4}{m0^2 \cdot c^4}\right) \cdot \left(\frac{1}{4 \cdot \pi \cdot \varepsilon 0}\right)^2 \cdot \frac{\left(1 + \cos\left(2 \cdot \theta_i\right)^2\right)}{2 \cdot \sin\left(2 \cdot \theta_i\right)} \cdot FF'_i \cdot \left(\lambda_i\right)^3 \cdot \frac{BArea \cdot dx}{va^2} \cdot e^{-\mu st_i \cdot \rho s \cdot (L - x)} \, dx \qquad (14)$$

Move all terms which are not a function of n outside the sum.

$$E2_i := I1_i \cdot \left[\left(\frac{e0^4}{m0^2 \cdot c^4}\right) \cdot \left(\frac{1}{4 \cdot \pi \cdot \varepsilon 0}\right)^2 \cdot \frac{\left(1 + \cos\left(2 \cdot \theta_i\right)^2\right)}{2 \cdot \sin\left(2 \cdot \theta_i\right)} \cdot FF'_i \cdot \left(\lambda_i\right)^3 \cdot \frac{BArea}{va^2}\right] \cdot \int_0^L \frac{e^{(-\mu st_i \cdot \rho s) \cdot x}}{\omega} \cdot e^{-\mu st_i \cdot \rho s \cdot (L - x)} \, dx$$

Solve the integral

$$E2_i := \frac{1}{32} \cdot I1_i \cdot \frac{e0^4}{\left[m0^2 \cdot \left[c^4 \cdot \left(\pi^2 \cdot \varepsilon 0^2\right)\right]\right]} \cdot \frac{\left(1 + \cos\left(2 \cdot \theta_i\right)^2\right)}{\sin\left(2 \cdot \theta_i\right)} \cdot FF'_i \cdot \left(\lambda_i\right)^3 \cdot \frac{BArea}{va^2} \cdot L \cdot \frac{\exp\left(-\mu st_i \cdot \rho s \cdot L\right)}{\omega} \qquad (15)$$

135

Considers Sample & Entrance Mold -->

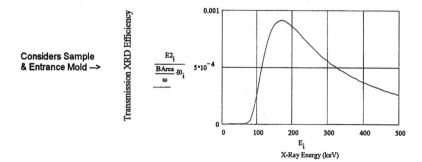

Considers Sample Only ----------------->

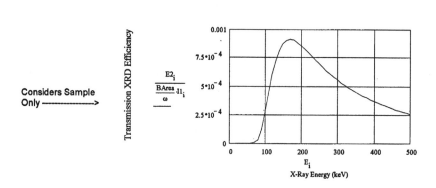

The diffracted energy is attenuated as it passes through the exit mold wall.

$$E3_i := E2_i \cdot \exp\left(- \mu wt_i \cdot \rho w \cdot W2\right) \tag{16}$$

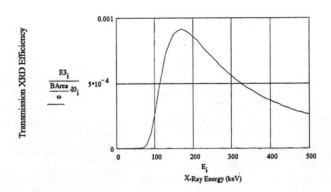

$$\text{WRITEPRN}("pfni10.dat") := \frac{E3_i}{BArea \cdot I0_i}$$

Calculate the intensity in the primary beam transmitted through both mold walls and the sample. The transmitted primary beam is what would normally be used to form a radiographic image.

$$It_i := I0_i \cdot \exp\left(- \mu wt_i \cdot \rho w \cdot W1\right) \cdot \exp\left(- \mu st_i \cdot \rho s \cdot L\right) \cdot \exp\left(- \mu wt_i \cdot \rho w \cdot W2\right) \tag{17}$$

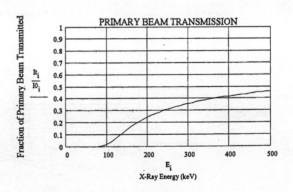

The plots are for a 0 mm entrance mold, 10 mm Ni sample, and a 0 mm exit mold.

137

US005589690A

United States Patent [19]

Siewert et al.

[11]	Patent Number:	**5,589,690**
[45]	Date of Patent:	**Dec. 31, 1996**

[54] **APPARATUS AND METHOD FOR MONITORING CASTING PROCESS**

[75] Inventors: **Thomas A. Siewert**. Boulder: **William P. Dubé**, Denver: **Dale W. Fitting**, Golden. all of Colo.

[73] Assignee: **National Institute of Standards and Technology**, Washington. D.C.

[21] Appl. No.: **407,699**

[22] Filed: **Mar. 21, 1995**

[51] Int. Cl.⁰ **G01N 23/20**
[52] U.S. Cl. **250/390.06**; 250/390.09; 378/71
[58] Field of Search 378/70, 73, 86, 378/71, 6; 250/390.09, 390.06

[56] **References Cited**

U.S. PATENT DOCUMENTS

3.499.736	3/1970	Zwaneburg	23/301
3.790.252	2/1974	Pao	350/160 R
3.833.810	9/1974	Efanov et al.	250/273
4.284.887	8/1981	Kusumoto et al.	250/272
4.634.490	1/1987	Tatsumi et al.	156/601
4.696.024	9/1987	Pesch	378/71
4.710.259	12/1987	Howe et al.	156/601
5.016.266	5/1991	Meurtin	378/73
5.093.573	3/1992	Mikoshiba et al.	250/310

5.136.624	8/1992	Schneider et al.	378/73
5.193.104	3/1993	Bastie et al.	378/73

FOREIGN PATENT DOCUMENTS

642638	1/1979	U.S.S.R.	378/73

OTHER PUBLICATIONS

"X–Ray Diffraction". Physical Metallurgy Principles by Reed-Hill, pp. 26–33.

Baxter et al., "Rapid Orientation Measurement of Single Crystal Casting Using 'Scorpio'". Insight. vol. 36. No. 5. May 1994.

Hashizume et al, "Techniques for Time–Resolved X–Ray Diffraction Using a Position Sensitive Counter". Japanese J. Appl. Phys., 15, (11), Nov. 1976. pp. 2211–2219.

Primary Examiner—Carolyn E. Fields
Attorney, Agent. or Firm—Sheridan Ross & McIntosh

[57] **ABSTRACT**

The present invention uses a high energy x-ray. neutron. or gamma source for monitoring the interface between a molten and solidified crystalline phase while in a furnace in a casting process. The radiation can also be used to determine the quality and orientation of the crystals in the crystalline phase. The invention uses the distinctive diffraction patterns produced by crystalline and amorphous phases to locate the interface.

25 Claims, 4 Drawing Sheets

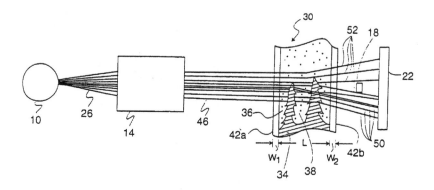

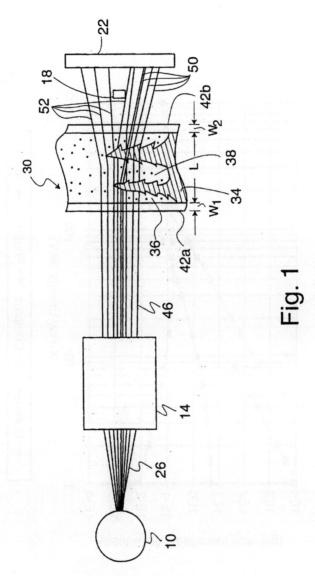

Fig. 1

139

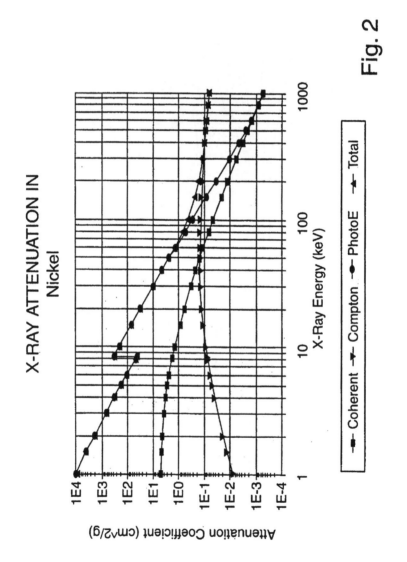

X-RAY ATTENUATION IN
Nickel

Fig. 2

140

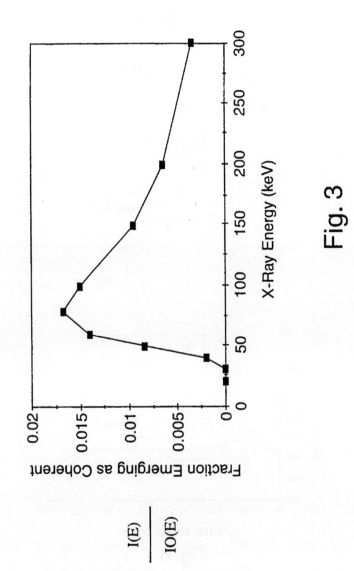

Fig. 3

141

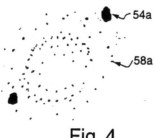

Fig. 4

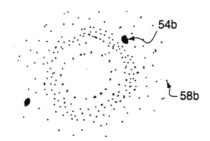

Fig. 5

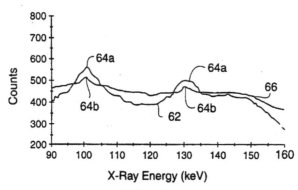

Fig. 6

142

1

APPARATUS AND METHOD FOR MONITORING CASTING PROCESS

FIELD OF THE INVENTION

The present invention relates generally to an apparatus and method for monitoring the casting or thermal treatment of materials, and more particularly, for monitoring the liquid amorphous/solid crystalline interface in a material being cast so as to permit process control parameters to be established and/or controlled in response thereto.

BACKGROUND OF THE INVENTION

Advancements in materials processing technology have caused an increasing demand for parts made from superalloys. Many high performance aircraft, for example, now employ turbine blades that are single crystals of nickel-based superalloys. The desirability of superalloys results from their high strength at high operating temperatures.

Superalloy parts are typically produced as single-crystal or directionally-solidified castings. Elaborate casting and inspection methods are employed to ensure that each part is a single-crystal or directionally-solidified crystal with a desired orientation. Even minor defects in the crystalline structure may be unacceptable as they can result in mechanical failure.

To reduce the likelihood of defects in the crystalline structure, the casting process for such parts has become a labor and time intensive process. The molten alloy is poured into a mold located in a furnace. One end of the mold is cooled to initiate crystallization. The mold is then slowly withdrawn from the furnace. The withdrawal rate is extremely slow to ensure an acceptable crystallization rate and crystal of the correct orientation. A slow crystallization rate promotes flawless crystal growth in the direction of solidification. If the rate of crystallization is too rapid, the metal will form unacceptable polycrystals and the part must be discarded. The possibility of forming parts with defects in the crystal structure with a lack of a reliable technique to otherwise monitor the rate of crystal growth during casting has to date caused engineers to select extremely conservative (i.e., slow) crystallization rates.

Further, despite the noted precautions taken during casting, defects can still occur which result in discarded parts and wasted production time. The scrapped material cannot simply be remelted and reused but must rather be retreated by an expensive refinement process before further use.

SUMMARY OF THE INVENTION

It is an objective of the present invention to provide an apparatus and method for real time monitoring of the interface between the solid crystalline and liquid amorphous phases of a material being cast in a mold during casting, thereby allowing for an increase in throughput and decrease in production costs. It is a related objective to provide a monitoring apparatus and method to permit casting process control parameters to be established and/or controlled in response thereto, wherein the process provides optimal conditions for the desired crystal formation.

It is a further specific objective to increase throughput and decrease the production costs of single-crystal and directionally-solidified castings by decreasing the time required to form such castings and minimizing the occurrence of defects and flaws in the castings. As will be appreciated, the present invention is particularly useful for monitoring the

interface between molten (i.e., liquified) and solidified states of materials, such as alloys of iron, titanium, nickel, and aluminum, ceramics, semiconductors, such as silicon, germanium, gallium arsenide, cadmium telluride, cadmium sulfide, and indium phosphide.

By way of initial summary, the present invention includes: (i) directing source radiation through a casting mold and into an interface between a portion of the crystalline phase and amorphous phase of a material being cast to provide diffracted radiation having a first component associated with the crystalline phase and a second component associated with the amorphous phase; (ii) receiving at least a portion of at least one and preferably both of the first component and second component of the diffracted radiation outside of tile casting mold and providing an output signal in response thereto; and (iii) using the output signal to monitor the interface between the crystalline and amorphous phases.

Preferably, x-ray radiation is employed and, as will be further discussed, the energy and intensity of the source radiation should be sufficient to penetrate the mold/cast material and to otherwise provide the diffracted components at an intensity sufficient for associated detection by a radiation-sensitive receiver means such as one or more x-ray imager(s) or energy detector(s). The casting mold can be composed of either amorphous or crystalline materials provided that the container is penetrable by the source radiation and stable at the melting point of the cast material. Suitable container materials include, for example, sand, alumina, zirconia, quartz, graphite, ceramics, and more particularly, aluminum oxide, zirconium oxide, and other metal oxides. Additional intervening structures, such as the walls of a furnace, can be present between the radiation source and diffracted radiation receiver, provided that such structures do not block or otherwise unduly attenuate the source and/or diffracted radiation.

The first and second components of the diffracted radiation result from the differing manners in which the amorphous phase and crystalline phase of the cast material diffract the source radiation. Briefly, the crystalline phase produces a radiation diffraction pattern comprising concentrated radiation areas, or high-intensity spots, while the amorphous phase produces a more diffuse and lower intensity ring pattern. Consequently, the radiation-sensitive receiver means should be positioned so as to receive at least a portion of one and preferably both of said concentrated radiation and diffuse radiation associated with the crystalline and amorphous phases, respectively.

In this manner, e.g., the output signal of the radiation-sensitive receiver means will reflect, on a real-time basis, an increase over time in the received diffracted radiation associated with an increasing crystalline phase. More preferably, e.g., by receiving a portion of both the diffracted radiation associated with an increasing crystalline phase and the diffracted radiation associated with the decreasing amorphous phase, the output signal can reflect the increase of the diffracted radiation received from the crystalline phase relative to that received from the amorphous phase. In this regard, the output signal may include a first output signal component corresponding with the first diffracted radiation component that is received and a second output signal component corresponding with the second diffracted radiation component that is received.

As will be appreciated then, use of the output signal to monitor the crystalline phase/amorphous phase interface generally entails (i) generating for successive time intervals corresponding successive values corresponding with at least one or preferably both of a first output signal component corresponding to the first diffracted radiation component and a second output signal component corresponding to the second diffracted radiation component; and (ii) employing the successive values to monitor the state and/or progression of the phase interface. More particularly, the successive values can be compared to predetermined completion values that correspond with associated degrees of crystallization completion, as determined by prior design testing.

For example, to determine the position of the interface as a function of time, the process can include the steps: (i) comparing during a first time interval the first and second output signal components to locate the interface at a first position; (ii) comparing during a second time interval the first and second output signal components to locate the interface at a second position; and (iii) employing the first and second positions to monitor the phase interface. The position of the interface can be estimated based upon the magnitude of the first and/or second output signal components. For example, the magnitude of the first output signal component can be compared to a predetermined magnitude corresponding with the diffraction pattern of a complete crystalline phase in the mold and/or the magnitude of the second output component can be compared to a predetermined magnitude corresponding with the diffraction pattern of a complete amorphous phase in the mold to locate the point below which the material in the container is substantially crystalline or above which the material in the container is substantially amorphous.

Further, the above-noted successive values can be successively compared to either predetermined completion values or to each other to determine a rate of crystallization. Such determination may include, e.g., (i) comparing during a first time interval a first output signal component (corresponding with the received portion of the first component of the diffracted radiation) with a second output signal component (corresponding to the received portion of the second signal component of the diffracted radiation) to provide a first compared value; (ii) comparing during a second time interval the first output signal component with the second output signal component to provide a second compared value; and (iii) employing the first and second compared values to determine a rate of crystallization. The rate of crystallization can be compared to predetermined rate values to monitor when crystallization is complete (e.g., as rate approaches zero) and/or to otherwise monitor if crystallization is occurring at a desired rate (e.g., to ensure the desired crystal structure).

As will be appreciated, by monitoring the crystalline phase/amorphous phase interface, a monitor signal can be generated for real-time control of at least one of the following casting process parameters: the temperature of a furnace containing the mold and the rate of withdrawal of the cast material from the furnace.

The present invention can be used not only to provide real-time monitoring and control of a casting process but also to verify in the laboratory casting process models leading to the establishment of casting process control parameters to yield substantially optimal conditions for crystal formation. The performance of casting under substantially optimal conditions reduces both the frequency of defects and/or flaws in the crystal structure of the part and the labor and time requirements to produce the part. These reductions substantially reduce the production costs of the parts relative to existing casting processes.

BRIEF DESCRIPTION OF THE DRAWINGS

FIG. 1 illustrates one embodiment according of the present invention;

FIG. 2 is a plot of the mass attenuation coefficients of nickel versus the energy of the source radiation;

FIG. 3 is a plot of the fraction of diffracted radiation emerging from a mold as coherent radiation as a function of the energy of source radiation;

FIGS. 4 and 5 depict diffraction patterns incident upon an x-ray imager for a sample containing crystalline and amorphous phases; and

FIG. 6 is a plot of the intensity detected by an energy detector versus the energy of the source radiation at two time intervals for a sample containing crystalline and amorphous phases.

DETAILED DESCRIPTION

The present invention provides an apparatus and method for the real-time monitoring of the interface between a crystalline phase, such as a solidified metal, and an amorphous phase, such as a molten metal, while contained in a mold. The interface is monitored by comparing the distinctly different radiation diffraction patterns produced by a crystalline phase compared to an amorphous phase. As discussed in detail below, the highly ordered crystal structure of the crystalline phase will produce a radiation diffraction pattern that includes a number of distinct spots and energy peaks while the relatively disordered structure of an amorphous phase will produce a radiation diffraction pattern that is a diffuse ring pattern with no distinct spots or energy peaks. Thus, the two phases produce radiation diffraction patterns that are distinct in geometry, intensity, and energy distribution. Although materials in the mold can produce interfering spots or rings, the radiation diffraction pattern produced by the mold is known, or is measurable, and can therefore be removed from or otherwise accounted for in the measured diffraction pattern.

FIG. 1 depicts an embodiment according to the present invention. The embodiment includes an x-ray radiation source 10, a collimator 14, a beam stop 18, and a detector assembly 22. The radiation source 10 produces x-ray source radiation 26 of a relatively high energy and intensity to penetrate a sample section 30, which includes the crystalline and amorphous phases 34, 38, and mold walls 42a,b. The collimator 14 restricts the source radiation 26 to a narrow range of angular orientations. This collimated radiation 46 has a plurality of substantially parallel x-rays. The beam stop 18 absorbs undiffracted and other radiation that has passed through the mold walls 42a,b and sample section 30 within a predetermined angular or spatial range. The detector assembly 22 receives a first component 50 of the diffracted radiation associated with the crystalline phase and a second component 52 associated with the amorphous phase and provides an output signal related to the first and second components.

The radiation source 10 can be any suitable x-ray, neutron, or gamma source for producing radiation having energy levels sufficient to penetrate the mold walls 42a,b, sample section 30, and any other intervening structure and a sufficient intensity to provide a diffraction pattern from the sample section 30 at the detector assembly 22 that is distinguishable from other incident radiation, such as non-diffracted radiation not absorbed by beam stop 18, leakage from the radiation source 10, and radiation diffracted by intervening structures, such as the mold walls 42a, 42b. More particularly, the energy level of radiation source 10 should be selected so that the collimated radiation 46 will travel through the first mold wall 42a, and a portion of the

5

sample section **30** so as to permit at least a portion of the radiation to undergo a coherent interaction with the crystalline phase **34** and/or amorphous phase. Further, the first and second components of the diffracted radiation (e.g., coherent radiation) must then penetrate the remainder of the sample section and the second mold wall **42b**.

In selecting the energy of the radiation source **10**, there is a tradeoff between achieving adequate transmission of the radiation through the mold walls **42a,b** and other intervening surfaces and the sample section **30** (the degree of transmission of radiation is greater at higher energies and lower at lower energies) and obtaining a sufficiently large cross-section for the first and second components of the diffracted radiation (the probability of coherent interactions,and therefore the quality of the diffraction image produced by the detector assembly is higher at lower energies than higher energies). At excessively high energies, Compton scattering of the radiation, which is substantially forward directed, can be a problem.

In selecting the intensity of the radiation source, the diffracted radiation received by the detector assembly **22** must have sufficient intensity to be separated from background noise. Background noise is typically caused by radiation diffracted by intervening structures, such as the mold, leakage radiation from the radiation source, Compton scatter from the sample and structures surrounding the sample. (e.g., mold, furnace), and non-diffracted radiation that is not fully absorbed by the beam stop **18**. It is necessary to select the radiation source intensity such that the intensity of the diffracted pattern exceeds the intensity of the background noise.

In view of the foregoing, in selecting both the energy and intensity of the radiation source **10**, the thicknesses and compositions of the sample section **30** and mold walls **42a,b** and other intervening structures (e.g., furnace walls) can be measured and otherwise accounted for. These properties are important for predicting losses in the structures.

Relatedly, the total mass attenuation coefficients for the sample section **30**, mold **42**, and other intervening structures can be determined for a predetermined range of source radiation energies utilizing known values. By way of example, FIG. **2** sets forth partial and total mass attenuation coefficients for nickel. As can be seen from FIG. **2**, the mass attenuation coefficients are a function of the source radiation energy.

After determining the total mass attenuation coefficients for the sample section **30**, mold **42**, and other intervening structures and the coefficient for coherent interactions in the sample at different potential energies of the source radiation, and the densities of the sample section **30**, mold **42**, and other intervening structures, a ratio between the intensity of the coherent radiation that would emerge from the sample section **30**, mold **42** and other intervening structures present (I(E)), and the intensity of the source radiation received by the detector assembly **22** in the absence of the sample section **30**, mold **42**, and other intervening structures (IO(E)) can be determined for each potential energy (E) under consideration for the source radiation. In this regard, a model of the transmission of radiation through a material encased in a mold can be represented by the following equation:

$$I(E)=IO(E)\cdot exp(-\mu wt(E)\cdot \rho w\cdot W1)\cdot NL\cdot (exp(-\mu st(E)\cdot \rho s\cdot L) -exp(-\rho s\cdot (\mu st(E)\cdot L+\mu sr(E)\cdot dx)))\cdot exp(-\mu wt(E)\cdot \rho w\cdot W2);$$

wherein for a given source intensity at the energy E:,

6

IO(E) is the intensity at the detector of the radiation beam, with no sample or mold present;

I(E) is the intensity of coherent (e.g., Rayleigh) scatter radiation which emerges from the sample section enclosed by mold walls;

$\mu wt(E)$ is the total mass attenuation coefficient in the mold material (cm^2 /g);

ρw is the density of the mold material (g/cm^3);

W1 is the thickness of the first mold wall **42a** (cm);

W2 is the thickness of the second mold wall **42b** (cm);

L is the thickness of the sample section **30**(cm);

NL is a large integer (e.g., 1000) which partitions the sample length (L) into small segments (dx);

$\mu st(E)$ is the total mass attenuation coefficient in the sample section **30** (cm^2/g);

ρs is the density of the sample section **30** (g/cm^3);

$\mu sr(E)$ is the mass attenuation coefficient for coherent (e.g., Rayleigh) radiation scattering in the sample section **30** (cm^2/g).

Additional intervening structures, such as a furnace wall, can be accounted for in the equation by adding exponential functions based on the total mass attenuation coefficient of the intervening structure material at energy E, the density of the material, and the thickness of the material. A more complex model could further include the spatial redistribution of radiation during diffraction.

The various ratios of I(E) to IO(E) can then be plotted as a function of the corresponding potential energies of the source radiation to identify the preferred mean energy of the source radiation. By way of example, FIG. **3** provides such a plot based on an aluminum sample (25 mm thick) encased in an aluminum oxide mold (with 6 mm thick walls). The peak in the plotted curve shows that there is an optimal range of source radiation energies. The optimum energy yields the highest production of diffracted radiation which is able to penetrate through the mold walls **42a,b** and sample section **30**. Because the mean energy of an x-ray source is generally one-half to one-third of the rated x-ray tube voltage the source preferably uses a tube voltage that is about two to three times the preferred energy peak.

The preferred intensity of the source radiation received by the detector assembly **22** is based on a predetermined minimum detectable radiation level for the particular detector assembly **22** used, the dynamic range of the detector assembly **22**, and the ambient noise radiation levels at the detector assembly **22** during operation. As noted above, the intensity of the diffracted radiation received by the detector assembly **22** must be sufficient to permit the diffracted radiation to be distinguished from the noise. As will be appreciated then, the tube current in the radiation source **10** should be selected to achieve the desired intensity at detector assembly **22**.

In this regard, the radiation source for diffraction spot imaging preferably is not a microfocus radiation source. In other words, the radiation source preferably has focal spot dimensions larger than about 0.1 mm. Microfocus radiation sources, though useful for thin samples and thin mold walls, have insufficient intensity to provide a detectable transmission diffraction pattern image for the thicknesses of the samples and mold walls normally encountered in casting processes. Microfocus radiation sources typically handle only small electrical currents (e.g., less than about 1 mA) which are often too low to produce the desired intensity of the source radiation for rapid inspection.

As noted, the collimator **14** collimates the divergent source radiation **26** to a small spatial size. Collimators for radiation sources generally are designed using several apertures (in radiation-attenuating material) placed along the source radiation beam path. The aperture dimensions of the collimator are a tradeoff between maximizing the intensity of the radiation beam transmitted (larger aperture size) and minimizing the size of the spatial resolution cell. For example, a radiation beam diameter of 1 mm has been found suitable for a radiation source-to-sample distance of 200 mm (e.g., 0.29 degrees angular aperture). Instead of collimating apertures, the collimator could be comprised of x-ray optic devices which restrict the radiation beam to the desired dimensions. The collimated radiation 46 contains a plurality of substantially parallel rays.

The beam stop 18 is an object composed of a material capable of absorbing a substantial portion of the transmitted (undiffracted) source radiation. A common type of beam stop 18 is composed of tungsten or lead.

The detector assembly 22 may comprise an x-ray imager or merely an energy-sensor, or a combination thereof. An x-ray imager allows for recording/processing of the spatial location of diffracted radiation by measuring the intensity of radiation over a two-dimensional area. An energy-sensitive detector allows for the recording/processing of the diffracted radiation received across the source energy spectrum.

FIGS. 4–5 illustrate the different diffraction patterns received at an x-ray imager for crystalline and amorphous material phases. The x-ray imager shows diffracted radiation from a crystalline phase (e.g., the first diffraction component) as a few spots, and from an amorphous phase (e.g., the second component) as a diffuse ring pattern. Specifically, FIG. 4 has spots 54a representing the first component and a diffuse ring pattern 58a representing the second component. FIG. 5 has less intense spots 54b from the first component and a more intense diffuse ring pattern 58b from the second component, indicating that the sample section examined in FIG. 4 has a greater volume of crystalline phase and a smaller volume of amorphous phase than the sample section examined in FIG. 5. The x and y axes of FIGS. 4 and 5 reflect spatial coordinates of the diffracted radiation received by the imager.

Referring to FIG. 6, the output of an energy sensitive detector is depicted versus the energy of the source radiation at two time intervals. More particularly, FIG. 6 plots the quantity of photons (or counts) interacting with the energy detector during a given time period versus the energy of the photons (i.e., corresponding with the radiation source energy spectrum). Peaks 64a,b correspond to the spots 54a,b, respectively, of FIGS. 4–5. The height of the peaks corresponds to the first component and thus to the amount of crystalline phase in the sample. Accordingly, curve 62 corresponds to a greater amount of crystalline phase in the sample than curve 66.

The unique Laue diffraction pattern of crystalline phases relative to amorphous phases is explained by the Bragg equation. The spots represent diffracted radiation that is a solution to the Bragg equation:

$$\frac{1.24n}{E} = n\lambda = 2d\sin\theta$$

where,

n is an integer;

E is the energy (keV) of the source radiation which is diffracted;

λ is the wavelength (nm) of the incident source radiation which is diffracted;

d is the lattice spacing (nm) in the crystal structure of the crystalline phase.

θ is the angle between the lattice plane and the incident radiation.

For a given orientation of the crystalline phase, diffracted radiation that is a solution to the Bragg equation forms spots 54a,b on the diffraction pattern. In contrast, the disordering of the amorphous phase produces the diffuse ring pattern 58a,b.

A detector assembly 22 utilizing the x-ray imager has different components from a detector assembly utilizing an energy-sensitive detector. A detector assembly 22 utilizing an x-ray imager generally includes a scintillator, an image intensifier, and a charge coupled device video imager. A detector assembly 22 utilizing an energy-sensitive detector includes the detector (which is preferably a cooled, intrinsic germanium detector with an integral FET preamplifier), a high voltage detector bias supply, a linear spectroscopy amplifier, and a multichannel analyzer. The components of the detector assembly preferably are optimized for high sensitivity in the 40 to about 500 keV range with a high count rate (e.g., greater than about 100,000 counts/sec) capability. Energy-sensitive detectors generally provide a higher sensitivity than the x-ray imager and thus require a lower intensity x-ray source.

The monitoring system in FIG. 1 is set to accommodate a transmission-type diffraction pattern. As will be appreciated, the system could also be arranged to accommodate a back reflection-type diffraction pattern in which the diffraction pattern is received by a detector assembly located on the radiation source side of the sample, or to accommodate a glancing incidence-type diffraction pattern in which the diffraction pattern is received by the detector assembly located at the side of the sample. The transmission-type Laue diffraction pattern is preferred as the diffraction patterns produced by the other two configurations are produced by diffracted radiation that may not have passed completely through the sample. It is important for the pattern to be produced by radiation passing completely through the sample to accurately locate the interface between the crystalline and amorphous phases at all locations within the casting cross-section. Additionally, the transmission mode is necessary to assess the relative crystalline/amorphous phase fractions along the radiation beam path.

Referring again to FIG. 1, the operation of a casting process using the present invention will now be described. The description will be based upon the use of the invention in the casting of single-crystal or directionally-solidified parts.

The casting process is commenced by placing the molten material into the mold 42 which rests on a chilled surface (not shown). The mold 42 is contained within an induction or resistance-heated furnace (not shown) and is slowly withdrawn from the furnace as crystallization progresses upwards from the chilled surface.

Source radiation 26 is generated by the radiation source 10, restricted spatially by the collimator 14 and directed through the mold wall 42a, a portion of the amorphous phase 38, the interface between the amorphous and crystalline phases, and a portion of the crystalline phase 34. The source radiation interacts with the crystalline and amorphous phases in the sample section to produce diffracted radiation having the first component 50 associated with the crystalline phase and the second component 52 associated with the amorphous phase.

At least a portion of the first and second components are received by the detector assembly 22 which provides an output signal related to the received portion of one and preferably each of the first and second components. The output signal is used to monitor the interface between the crystalline and amorphous phases.

The output signal can be used to estimate the position of the interface in the mold and/or the degree of crystallization of the crystalline phase. To do this, the first component may be compared to the second component at selected time intervals. For example, at a first time interval the source radiation may be scanned vertically up and down the mold to locate either (i) the interface between the portion of the sample that is entirely an amorphous phase and that which has both amorphous and crystalline phases (e.g., the top of the dendrites **36**) and/or (ii) the interface between the portion of the sample that is entirely a crystalline phase and that which has both amorphous and crystalline phases (e.g., the base of the dendrites **36**). At a second time interval, the above-noted interfaces are again located. The distance between the interface at the first time interval and the interface at the second time interval divided by the period between the first and second time intervals provides the rate of crystallization.

The first and second components can also be measured for a defined area of the sample during different time intervals to estimate the rate of crystallization. For example, the first and second components are compared with one another during a first time interval to determine a first degree of crystallization of the crystalline phase and during a second time interval to determine a second degree of crystallization of the crystalline phase. The comparison can be based on the relative intensities or magnitudes of the first and second components. The crystallization rate can be determined based on the first and second degrees of crystallization and the period between the time intervals.

As will be appreciated, computed tomography can be applied to data regarding the position and/or crystallization rate to generate a two- or three-dimensional representation of the interface between the amorphous and crystalline phases. Computed tomography using the first and second components of diffraction will yield tomography images with higher contrast between crystalline and amorphous phases than computed tomographic images based on exploiting the small density differences between the crystalline and amorphous phases. Such an image could be generated during discrete time intervals to more precisely control the casting process or to verify casting-process models.

Based on the position and/or crystallization rate, a monitor signal can be provided to control the casting process. The monitor signal can be used to control selected parameters in the casting process, such as the temperature of the furnace and/or the rate of withdrawal of the mold from the furnace (e.g., the duration of the casting), and the like. The monitor signal can be manually or automatically generated. To automatically generate the monitor signal, the first and second components or the interface position or crystallization rate can be compared against predetermined values corresponding to each time interval during which the measurements are taken. If the measurement exceeds a maximum predetermined value (e.g., the rate of crystallization is too fast) or is less than a minimum predetermined value (e.g., the rate of crystallization is too slow), an appropriate monitor signal is generated to adjust the casting process.

To form the desired diffraction patterns, the source radiation should be collimated. Preferably, the intensity of the portion of the first component received by the detector assembly and the portion of the second component received by the detector assembly together is at least about 0.1% of the intensity of the source radiation.

The monitor signal can be used to cause a portion of the mold to be heated to a second temperature greater than a first

temperature of the mold to remelt a portion of the crystalline phase. After the portion of the crystalline phase is remelted, the portion of the mold can be cooled to a third temperature less than the second temperature to recrystallize the remelted portion of the crystalline phase. It is believed that the remelting and recrystallization of the defective portions of the crystalline phase is an effective solution to decrease the number of castings which must be discarded for crystalline defects.

As will be appreciated, the present invention is not necessarily limited to casting processes but can be used in any process to monitor the position of the interface between a crystalline phase and an amorphous phase, such as a liquid, or any other process involving an interface between a crystalline and another crystalline or an amorphous phase. The present invention can thus be used to monitor solidification processes, such as the casting of metals and alloys, growth of semiconductor boules, chemical precipitation processes, such as those in which crystalline precipitates are formed at high temperatures in an amorphous phase.

By way of example, the present invention can detect the formation of a second crystalline phase in a crystalline structure. The second crystalline phase has a different crystal structure than the surrounding crystal. The formation of a second crystalline phase in a surrounding crystalline phase is employed in the alloying of metals. The formation of the second crystalline phase would produce diffracted radiation different from the diffracted radiation produced by the surrounding crystalline phase in the absence of the second crystalline phase. The diffraction pattern of the second crystalline phase can be isolated by taking into account the diffraction pattern produced by the surrounding crystalline phase in the absence of the second crystalline phase.

Additionally, the present invention can be used to detect defects in the crystal structure of the crystalline phase during casting. A material having multiple crystals will produce a diffraction pattern distinct from a crystalline phase having only one crystal. For example, a crystalline having multiple crystals will produce more spots than a crystalline phase having only one crystal. In the event that more than one crystal is detected in a monocrystalline phase would enable the casting process to be altered to remelt the portion of the material having more than one crystalline structure followed by resolidification of the crystalline phase. In this matter, the portion of the castings discarded for crystalline defects can be significantly reduced.

EXPERIMENT 1

A small piece of directionally-solidified turbine blade alloy was polished to make the directional nature of the microstructure observable. Incident x-ray beam and film plane angles were adjusted to 45° to give an angle of 90° between the x-ray beam and imager. A definite x-ray diffraction pattern was obtained using a 50 kVp x-ray tube voltage. The x-ray tube voltage was increased to 75 kVp, then to 100 kVp and finally to 150 kVp. The diffraction pattern persisted, with a very noticeable increase in intensity. The Compton scattering background increased somewhat with tube voltage increases. This experiment verifies that x-ray diffraction can be obtained with x-ray tube voltages (100 kVp and 150 kVp) which are substantially higher than that used in conventional (40 kVp) Laue measurements.

EXPERIMENT 2

Experiment 1 was repeated with a 6 mm-thick piece of mold material typically used for single-crystal turbine blade

11

castings (a mixture of aluminum oxide, silicon oxide, and zirconium oxide) in a position between the source radiation and the turbine-blade alloy specimen. At an x-ray tube voltage of 150 kVp, many of the diffraction spots near the beam stop (small-angle diffraction) remained. The experiment confirms the ability to perform transmission diffraction even through typical casting mold material.

EXPERIMENT 3

A series of heating, melting, and recrystallization experiments were conducted using x-rays to locate the advancing and receding interface between the solidified metal and molten metal. A 22-millimeter-diameter polycrystalline 99.999% aluminum rod was placed in a quartz tube in a furnace. The rod was heated to (652° C.) (near the melting point) and the changing transmission Laue diffraction pattern was observed. The specimen was then cooled and sectioned for analysis.

Another experiment was conducted in which the aluminum rod was melted and resolidified. Real-time x-ray diffraction was employed to follow the progression of the interface between the solidified aluminum and the molten aluminum. The difference between the pattern generated by the high-temperature solid aluminum and the pattern generated by the liquid aluminum was dramatic and unmistakable. The diffraction spots from the solid disappeared as a diffuse ring formed when the aluminum was fully melted.

EXPERIMENT 4

A copper rod was placed in a quartz tube, with a triangular cross-section. The tube was then inserted into a gradient furnace. A 1 millimeter diameter, collimated x-ray beam was directed into the furnace, and a real-time x-ray imager was placed on the opposite side of the furnace. A 6 millimeter thick, 3 millimeter diameter tungsten disk was positioned in the center of the primary x-ray beam emerging from the furnace to act as a beam stop. X-rays diffracted from the copper sample in the furnace passed to the sides of the beam stop and were imaged.

The furnace was manipulated remotely to move it vertically and horizontally. The vertical movement was used to scan the x-ray beam and imager with respect to the liquid/solid copper boundary that was established in the gradient furnace. Horizontal movement across the wedge-shaped copper specimen permitted interrogating different thicknesses.

The temperature of the furnace was raised to melt the copper and then lowered to solidify it. The warming and cooling sequence was repeated several times. As in the aluminum melting experiments, the solid copper produced a diffraction image with bright diffraction spots. When the melting temperature of copper was exceeded, the ordered diffraction pattern disappeared and was replaced by a diffuse ring of x-ray scattering from the molten copper. This experiment validated the sensing method for locating a liquid/solid boundary in a metal sample with physical characteristics (atomic number and density) similar to that of nickel-based alloys.

EXPERIMENT 5

A gallium sample was placed in a container capable of producing a temperature gradient. A temperature controller was connected to the heater to produce a steady-state boundary between the solid gallium (at the top because its

12

density is less than the liquid gallium) and the liquid gallium. The position of the probing x-ray beam was moved into the solid or liquid by a remote positioning fixture. X-ray diffraction images and radiographic images (by removing the collimating apertures in the x-ray beam) were alternatively taken. The density variation between the solid and liquid gallium is great enough to produce a discernable difference in brightness in a radiographic image. The ability to independently determine by radiographic image the position of the liquid/solid boundary (brighter versus darker regions of the radiograph) provided a means for validating the spatial performance of the x-ray imager.

When the x-ray imager sensor was positioned to probe only the solid gallium, bright diffraction spots were observed. In the liquid, the diffuse scattering ring was observed. At intermediate locations between liquid and solid, the diffuse ring and diffraction spots were present, but both with decreased intensity in rough ratio to their relative amounts. The radiographic images produced with the x-ray images validated that the x-ray imager could accurately locate the position of the liquid/solid interface, and verified that the spatial resolution of the diffraction sensor is approximately the size of the source radiation beam at the sample.

EXPERIMENT 6

The x-ray imager was replaced by an energy-sensitive detector (intrinsic germanium). The much higher efficiency of the detector, compared to the imager, required the replacement of the collimating apertures. The 1-millimeter diameter aperture, used in experiments with the x-ray imager, produced too intense a diffracted beam for the germanium detector. A 0.2 millimeter diameter aperture was used. A source-sample distance of 250 millimeters and a sample-detector spacing of 180 millimeters was used for the experiments. The spectra were obtained with an x-ray tube potential of 160 kV and a tube current of 1 mA.

Spectral peaks at approximately 100 keV and 130 keV were produced by x-ray diffraction spots in a solid gallium sample. There were no discernable peaks in the 100–130 keV spectrum recorded when the x-ray diffraction pattern in a liquid gallium specimen was examined. Although the intensity (x-rays per unit area per second) was higher for the diffraction spots, the spots were highly localized. The spectrum for the liquid specimen, in contrast, recorded more counts overall in each energy interval although the peak intensity was lower.

While various embodiments of the present invention have been described in detail, it is apparent that modifications and adaptations of those embodiments will occur to those skilled in the art. However, it is to be expressly understood that such modifications and adaptations are within the scope of the present invention, as set forth in the following claims.

What is claimed is:

1. A method for monitoring an interface between a crystalline phase and an amorphous phase of a material within a container, comprising the steps of:

directing source radiation through at least one wall of a container and into an interface between a portion of a crystalline phase of a material and a portion of an amorphous phase of said material contained in said container to provide diffracted radiation having a first diffraction component with a first spatial radiation distribution associated with said crystalline phase and a second diffraction component with a second spatial radiation distribution associated with said amorphous

148

13

phase, said first and second spatial radiation distributions being spatially distinct;

receiving at least a portion of at least one of said first diffraction component and said second diffraction component of said diffracted radiation on a radiation sensitive detector means located outside of said container to provide an output signal, said output signal including information corresponding to said spatial radiation distribution of the received portion; and

using said output signal to monitor said interface between said crystalline phase and said amorphous phase.

2. The method as, claimed in claim 1, wherein:

said amorphous phase is a liquid phase of said material.

3. The method as claimed in claim 2, wherein said material has a variable rate of crystallization, and said using step comprises:

comparing during a first time interval a first output signal component corresponding with said first diffraction component with a second output signal component corresponding with said second diffraction component to provide a first compared value;

comparing during a second time interval said first output signal component with said second output signal component to provide a second compared value; and

employing said first and second compared values to determine said rate of crystallization between said first and second time intervals.

4. The method as claimed in claim 1, wherein said receiving step comprises:

receiving both said first diffraction component and said second diffraction component on said radiation sensitive detector means.

5. The method as claimed in claim 4, wherein said container is a casting mold.

6. The method as claimed in claim 1, wherein said directing step comprises:

varying the intensities of said first and second diffraction components over time by changing the spatial location of said source radiation in a defined volume of said material.

7. The method as claimed in claim 1, wherein said receiving step comprises:

generating for successive time intervals corresponding successive values relating to at least one of a first output signal component corresponding with said first diffraction component and a second output signal component corresponding with said second diffraction component.

8. The method as claimed in claim 7, wherein said using step comprises:

comparing said successive values with a predetermined completion value that corresponds with a degree of crystallization completion.

9. The method as claimed in claim 7, wherein said using step comprises:

employing said successive values to determine a rate of crystallization.

10. The method as claimed in claim 9, wherein said using step comprises:

comparing said rate of crystallization to a predetermined value.

11. The method as claimed in claim 7, wherein said successive values corresponding to said first output signal component and said successive values increase in magnitude over said successive time intervals.

12. The method as claimed in claim 7, wherein said successive values corresponding to said second output signal

14

component and said successive values decrease in magnitude over said successive time intervals.

13. The method as claimed in claim 7, wherein:

said first and second output signal components correspond to the intensity of said first and second diffraction components, respectively.

14. The method as claimed in claim 1, wherein said source radiation is passed through a defined region of said container and said using step comprises:

comparing a first output signal component corresponding with said first diffraction component with a second output signal component corresponding with said second diffraction component to determine the portion of said defined region occupied by at least one of said crystalline phase and said amorphous phase.

15. The method as claimed in claim 1, wherein said using step comprises:

comparing during a first time interval a first output signal component corresponding with said first diffraction component and a second output signal component corresponding with said second diffraction component to locate said interface at a first position;

comparing during a second time interval said first and second output signal components to locate said interface at a second position; and

employing said first and second positions to determine a monitor signal.

16. The method as claimed in claim 15, wherein said first and second output signal components correspond to a time interval and said using step comprises:

comparing said output signal to a predetermined completion value.

17. The method as claimed in claim 1, wherein said receiving step comprises:

comparing said first and second spatial radiation distributions to provide said output signal.

18. The method as claimed in claim 17, wherein said using step comprises:

using said monitor signal to establish at least one casting process control parameter.

19. The method as claimed in claim 17, wherein said using step comprises:

using said monitor signal to control in real time at least one casting process control parameter.

20. The method as claimed in claim 17, further comprising:

using said monitor signal to control at least one of the following parameters: the temperature of a furnace containing said crystalline and amorphous phases and the rate of withdrawal of said crystalline phase from said furnace.

21. The method as claimed in claim 1, wherein a portion of said container is at a first temperature and further comprising:

heating in response to said output signal said portion of said container to a second temperature greater than said first temperature to remelt a portion of said crystalline phase.

22. The method as claimed in claim 21, further comprising:

cooling in response to said output signal said portion of said container to a third temperature less than said second temperature to recrystallize said remelted portion of said crystalline phase.

149

15

23. The method as claimed in claim **1**, wherein:

said source radiation and the radiation sensitive detector means are positioned on opposing sides of said material to produce transmission-type diffraction in said first and second diffraction components.

24. The method as claimed in claim **1**, wherein said material crystallizes in a substantially vertical direction and wherein:

said directing step comprises:

moving the source radiation relative to the container in a substantially vertical direction to monitor the progression of the interface over time.

16

25. The method as claimed in claim **1**, further comprising:

using at least one of a first output signal component corresponding with said first diffraction component and a second output signal component corresponding with said second diffraction component to perform computed tomography and generate a two- or three- dimensional representation of the interface between the crystalline and amorphous phases.

* * * * *

150

NIST *Technical Publications*

Periodical

Journal of Research of the National Institute of Standards and Technology—Reports NIST research and development in those disciplines of the physical and engineering sciences in which the Institute is active. These include physics, chemistry, engineering, mathematics, and computer sciences. Papers cover a broad range of subjects, with major emphasis on measurement methodology and the basic technology underlying standardization. Also included from time to time are survey articles on topics closely related to the Institute's technical and scientific programs. Issued six times a year.

Nonperiodicals

Monographs—Major contributions to the technical literature on various subjects related to the Institute's scientific and technical activities.

Handbooks—Recommended codes of engineering and industrial practice (including safety codes) developed in cooperation with interested industries, professional organizations, and regulatory bodies.

Special Publications—Include proceedings of conferences sponsored by NIST, NIST annual reports, and other special publications appropriate to this grouping such as wall charts, pocket cards, and bibliographies.

Applied Mathematics Series—Mathematical tables, manuals, and studies of special interest to physicists, engineers, chemists, biologists, mathematicians, computer programmers, and others engaged in scientific and technical work.

National Standard Reference Data Series—Provides quantitative data on the physical and chemical properties of materials, compiled from the world's literature and critically evaluated. Developed under a worldwide program coordinated by NIST under the authority of the National Standard Data Act (Public Law 90-396). NOTE: The Journal of Physical and Chemical Reference Data (JPCRD) is published bi-monthly for NIST by the American Chemical Society (ACS) and the American Institute of Physics (AIP). Subscriptions, reprints, and supplements are available from ACS, 1155 Sixteenth St., NW, Washington, DC 20056.

Building Science Series—Disseminates technical information developed at the Institute on building materials, components, systems, and whole structures. The series presents research results, test methods, and performance criteria related to the structural and environmental functions and the durability and safety characteristics of building elements and systems.

Technical Notes—Studies or reports which are complete in themselves but restrictive in their treatment of a subject. Analogous to monographs but not so comprehensive in scope or definitive in treatment of the subject area. Often serve as a vehicle for final reports of work performed at NIST under the sponsorship of other government agencies.

Voluntary Product Standards—Developed under procedures published by the Department of Commerce in Part 10, Title 15, of the Code of Federal Regulations. The standards establish nationally recognized requirements for products, and provide all concerned interests with a basis for common understanding of the characteristics of the products. NIST administers this program in support of the efforts of private-sector standardizing organizations.

Consumer Information Series—Practical information, based on NIST research and experience, covering areas of interest to the consumer. Easily understandable language and illustrations provide useful background knowledge for shopping in today's technological marketplace.
Order the above NIST publications from: Superintendent of Documents, Government Printing Office, Washington, DC 20402.
Order the following NIST publications—FIPS and NISTIRs—from the National Technical Information Service, Springfield, VA 22161.

Federal Information Processing Standards Publications (FIPS PUB)—Publications in this series collectively constitute the Federal Information Processing Standards Register. The Register serves as the official source of information in the Federal Government regarding standards issued by NIST pursuant to the Federal Property and Administrative Services Act of 1949 as amended, Public Law 89-306 (79 Stat. 1127), and as implemented by Executive Order 11717 (38 FR 12315, dated May 11, 1973) and Part 6 of Title 15 CFR (Code of Federal Regulations).

NIST Interagency Reports (NISTIR)—A special series of interim or final reports on work performed by NIST for outside sponsors (both government and non-government). In general, initial distribution is handled by the sponsor; public distribution is by the National Technical Information Service, Springfield, VA 22161, in paper copy or microfiche form.